A New Deal for Transport: Better for Everyone

The Government's White Paper on the Future of Transport

Presented to Parliament by the Deputy Prime Minister and
Secretary of State for the Environment, Transport and the Regions,
and the Secretaries of State for Scotland, Wales and Northern Ireland
by Command of Her Majesty
July 1998

Cm 3950 £16.50

Department of the Environment, Transport and the Regions
Eland House
Bressenden Place
London SW1E 5DU
Telephone: 0171 890 3000
Internet service http://www.detr.gov.uk/

© Crown copyright 1998

Copyright in the typographical arrangement and design vests in the Crown.

Published with the permission of the Department of the Environment, Transport and the
Regions on behalf of the Controller of Her Majesty's Stationery Office.

Extracts of this publication may be made for non-commercial in-house use, subject to the source being acknowledged.

Applications for reproduction should be made in writing to The Copyright Unit, Her Majesty's
Stationery Office, St Clements House, 2-16 Colegate, Norwich NR1 1BQ.

ISBN 0 10 139502 7

Printed in Great Britain on material containing 75% post-consumer waste and 25% ECF pulp.
July 1998.

Foreword

There is now a consensus for radical change in transport policy. The previous Government's Green Paper paved the way with recognition that we needed to improve public transport and reduce dependence on the car. Businesses, unions, environmental organisations and individuals throughout Britain share that analysis. This White Paper builds on that foundation.

For the last two decades, the ideology of privatisation, competition and deregulation has dominated transport policy. Bus and rail services have declined whilst traffic growth has resulted in more congestion and worsening pollution.

This White Paper fulfills our manifesto commitment to create a better, more integrated transport system to tackle the problems of congestion and pollution we have inherited. It is timely. In its Green Paper the previous Government recognised that we could not go on as before, building more and more new roads to accommodate the growth in car traffic. With our new obligations to meet targets on climate change, the need for a new approach is urgent.

As a car driver, I recognise that motorists will not readily switch to public transport unless it is significantly better and more reliable. The main aim of this White Paper is to increase personal choice by improving the alternatives and to secure mobility that is sustainable in the long term.

Better public transport will encourage more people to use it. But the car will remain important to the mobility of millions of people and the numbers of people owning cars will continue to grow. So we also want to make life better for the motorist. The priority will be maintaining existing roads rather than building new ones and better management of the road network to improve reliability.

More bus lanes, properly enforced, will make buses quicker and more reliable. Even a small increase in the numbers of bus passengers will transform the economics of the bus industry, allowing higher levels of investment in new buses and new and more frequent services.

This White Paper isn't just about national policy. *Local transport plans* will create a partnership between local councils, businesses, operators and users. Local initiatives such as safer routes to schools will give parents more confidence in letting their children make their own way. CCTV cameras in car parks and bus stations will make users, especially women, feel safer.

We have had to make hard choices on how to combat congestion and pollution while persuading people to use their cars a little less – and public transport a little more. And we have devised imaginative new ways of raising money from transport for better transport. That is the *New Deal for transport* which I believe the country wants.

The last transport White Paper was a generation ago. But the economy, technology and attitudes to transport and the environment are changing so rapidly that we should not wait another generation before a new White Paper. The new *Commission for Integrated Transport* will bring together transport users, the private sector, local authorities and others to make recommendations to Ministers.

This White Paper reflects the Government's commitment to giving transport the highest possible priority. We now look to others – companies, individuals, employees and local authorities – to join us in shaping a new future for sustainable transport in the UK.

JOHN PRESCOTT

CONTENTS
A New Deal for Transport

SCOPE 8

PART I

CHAPTER 1: A NEW DEAL FOR TRANSPORT 10
Lives shaped by transport
A new approach: A New Deal for transport

PART II

CHAPTER 2: SUSTAINABLE TRANSPORT 22
The New Deal for transport
Better health
More jobs and a strong economy
A better environment
A fairer, more inclusive society
A modern, integrated transport system
Changing travel habits
Technology taking the strain
The New Deal for transport – making a difference:
– on climate change
– on traffic and congestion
– on local air quality
– a more inclusive society
– through extending the range of targets

CHAPTER 3: INTEGRATED TRANSPORT 37

More choice 37
Making it easier to walk
Making it easier to cycle
More and better buses
A better railway
Better for the motorist
Moving freight
Getting to the airport
The role of motorcycling

More integrated public transport 47
In pursuit of the 'seamless' journey
Fares and ticketing
Physical interchange
Timetable co-ordination and service stability
Passenger information
Better taxis
Travelling without fear
Accessible transport for disabled people and easier access for all

Streets for people 58
Integration on local roads
Living town centres
Quality residential environments
A more peaceful countryside

Making better use of trunk roads 63
Integration
Investment strategy
An integrated network
A core road network
Making better use
The Highways Agency as network operator
Helping the road user
Better information for the driver
More care for the local environment
Better development control

Delivering the goods: sustainable distribution 70
Improving efficiency
Quality Partnerships for freight
Suitable traffic for suitable roads
Sustainable air freight
Sustainable shipping
Making better use of coastal shipping and inland waterways

Better integration of airports and ports — 76
- Integrated airports
- Integrated ports
- Trans-European Networks

Travelling safely — 82
- Road safety
- Railway safety
- Marine safety
- Air safety
- An integrated approach to transport safety

PART III

CHAPTER 4: MAKING IT HAPPEN — 92

European action — 92

UK action — 92
- Commission for Integrated Transport
- Funding transport
- Strategic Rail Authority
- Railways – fares
  – better services
  – the passenger's voice
- Rail Regulator – infrastructure investment
  – rolling stock leasing companies
- Investment in rail
- Investment in trunk roads
- Aviation and airport regulation
- Investment in aviation
- Trust ports
- Devolution

Regional action — 102
- Integrating transport and planning in the English regions
- Regional transport strategies
- Role of Regional Development Agencies
- Integrated transport in London
- Role of Passenger Transport Authorities

Local action — 111
- Local transport plans
- Funding bus services
- Reducing social exclusion
- Funding major local transport schemes
- Funding local rail services

Changing travel habits — 114
- Tackling congestion and pollution on local roads
- Charging users on motorways and trunk roads
- Workplace parking
- Non workplace parking

Sending the right signals — 119
- Economic instruments
- Cleaner, more efficient vehicles and fuels: fiscal incentives
- Company cars
- Incentives for green travel

Setting standards — 123
- Cleaner, more efficient vehicles and fuels: standards
- Better air quality
- Ports and shipping
- Air transport

Better planning — 126
- New policy guidance
- Planning guidance for – transport
  – housing
  – development plans
- Better implementation in the planning process
- Good design

Better enforcement — 129
- Better enforcement: road traffic
- Technology for enforcement
- Role of other agencies
- Police organisation
- British Transport Police
- Wheelclamping on private land
- Better enforcement: freight transport

Better appraisal	132	– journeys to work	
Transport impact assessment		– teleworking	
New approach to appraisal for transport projects		– school journeys	
		Building communities – community transport and rural transport partnership	
Economic appraisal		Raising awareness and informing choice	
Environmental appraisal		A new direction	
Improving appraisal: the planning process and development proposals			
Understanding the effects of noise		ANNEX A Future publications	154
Noise standards			
Noise mitigation		ANNEX B Consultation on integrated transport policy	155
Technology – research and development	137		
		ANNEX C Royal Commission on Environmental Pollution	156
CHAPTER 5: SHARING RESPONSIBILITY			
Partnership for action	139	ANNEX D 'Transport: The Way Forward'	158
Partnership in innovation and design			
Partnership: helping the motorist		ANNEX E Core trunk road network map	161
Working with transport operators			
Bus design		ANNEX F Rail network pinch-points	162
Working with business			
Local partnership		INDEX	00
A shared responsibility: individuals, families and communities:			

Acknowledgements:

Chapter 1: Congestion – courtesy of Alan Laughlin, City of Edinburgh Council

Chapter 2: Cyclist – courtesy of the Highways Agency

Chapter 3:
Artist's impression of Trafalgar Square – courtesy of Foster and Partners
Cycle lane, National Cycle Network map – ©SUSTRANS
Edinburgh Greenway bus lane – courtesy of Alan Laughlin, City of Edinburgh Council
Birkenhead bus station – courtesy of Merseytravel
Wheelchair user – ©GMPTE1998
Freight on inland waterway – courtesy of British Waterways Photolibrary

Luton Airport – courtesy of Luton Airport
Eurostar train – courtesy of Eurostar (IJK) Ltd.
'Piggyback' lorry – courtesy of Freight Transport Association
Northern Line Train – courtesy of London Transport
City of Edinburgh Council

Chapter 4: Vehicle Inspectors – courtesy of the Vehicle Inspectorate
Solar powered car – courtesy of Honda (UK)

Chapter 5: Smiling children – ©SUSTRANS

Annex E: Core trunk road network – map courtesy of the Highways Agency

Annex F: Rail network pinch-points – courtesy of Railtrack

Source of information for statistics, graphs etc is Department of the Environment, Transport and the Regions unless otherwise specified

SCOPE OF THE WHITE PAPER

This is a United Kingdom White Paper. It sets out a new approach to transport policy which has relevance throughout the United Kingdom, and it embodies new, modern thinking on integrating transport with other aspects of Government policy. Some of the discussion in the text relates only to England. But the guiding principles apply throughout the UK.

Different parts of the UK have differing transport needs. Scotland, Wales and Northern Ireland will be able to consider their own transport priorities under the new arrangements for a Scottish Parliament, a National Assembly for Wales and an Assembly for Northern Ireland. The Secretary of State for Scotland is publishing a White Paper on integrated transport policy in Scotland that sets out our transport policy for Scotland consistent with the principles in this paper. Separate documents will also be published for Wales and Northern Ireland.

In Northern Ireland, responsibilities that fall to local authorities in Great Britain for roads, transport, land use planning and the environment rest with the Department of the Environment for Northern Ireland and references to local authorities in this document should be read accordingly.

This White Paper sets the framework within which our detailed policies will be taken forward. Some of the proposals will require legislation which will be brought forward as soon as Parliamentary time allows.

A number of supporting documents which set out fuller details of the proposals highlighted in this White Paper will be published and are listed at Annex A. A summary of the responses to our consultation on integrated transport policy is at Annex B and a fuller summary is being published to accompany this White Paper.

PART 1

CHAPTER 1 A New Deal for Transport

A New Deal for transport: better for everyone

CHAPTER 1
A New Deal for Transport

> *"We will safeguard our environment, and develop an integrated transport policy to fight congestion and pollution."*
>
> Labour Party Manifesto, 1997

Lives shaped by transport

1.1 Our quality of life depends on transport. Most of us travel every day, even if only locally. And we need an efficient transport system to support a strong and prosperous economy. But in turn, the way we travel is damaging our towns and cities and harming our countryside. As demand for transport grows, we are even changing the very climate of our planet.

1.2 Cars in particular have revolutionised the way we live, bringing great flexibility and widening horizons. And we do not want to restrict car ownership – with our vision for a prosperous Britain where prosperity is shared by all we expect more people to be able to afford a car. But the way we are *using* our cars has a price – for our health, for the economy and for the environment[1].

1.3 Transport policies dominated by the short-term have reduced choice, for the public transport passenger and for motorists. The mood is for change. Business is concerned about the costs of congestion. People want the existing transport system to work better. They want more choice and a new emphasis on protecting the environment and their health.

1.4 Simply building more and more roads is not the answer to traffic growth. 'Predict and provide' didn't work. Privatisation and deregulation of public transport were key features of the last decade. But they failed the passenger because they fragmented public transport networks and ignored the public interest. This is why we promised an integrated transport policy to fight congestion and pollution.

1.5 In this White Paper, we set out our integrated transport policy. We explain how we will extend choice in transport and secure mobility in a way that supports sustainable development. It is our *New Deal for transport* – a transport system that is safe, efficient, clean and fair.

Congestion harms our environment and economy.

---

1 Information on the major trends in domestic transport is provided in "Transport Trends", Department of the Environment, Transport and the Regions, TSO, 1998. ISBN 0-11-551987-4.

Transport has enriched our lives but at a cost ....

1.6 Over 35 years ago, the Government of the day commissioned a study into the problems posed by road traffic. The resulting 'Buchanan report'[2] predicted that traffic would increase dramatically, with profound consequences for the environment and the way life was lived. It has. We cannot say that we weren't warned.

1.7 Congestion and unreliability of journeys add to the costs of business, undermining competitiveness particularly in our towns and cities where traffic is worst. The CBI has put the cost to the British economy at around £15 billion every year, some estimates are lower but agree that the cost to the nation runs into billions of pounds every year and is rising[3]. The convenience of the car is eroded by congestion and driving is increasingly stressful.

1.8 In the UK, emissions of CO_2 from road transport are the fastest growing contributor to climate change – the greatest global environmental threat facing the international community. Climate change doesn't mean we will all enjoy pleasant Mediterranean summers: it threatens unpredictable extremes of weather with more frequent and intense storms, floods, droughts and rising sea levels.

1.9 Road traffic is also adding substantially to the local air pollution that is damaging our health and hastens the death of thousands each year. Contrary to popular opinion, drivers and their passengers are not protected from the pollution they create – the air inside a car can be more polluted than for the pedestrian on the pavement.

.... which keeps going up

1.10 With increasing prosperity, more people with driving licences and several million new households likely over the next two decades, we are faced with dramatic increases in traffic. Over the next 20 years car traffic could grow by more than a third. Van and lorry traffic is forecast to grow even faster.

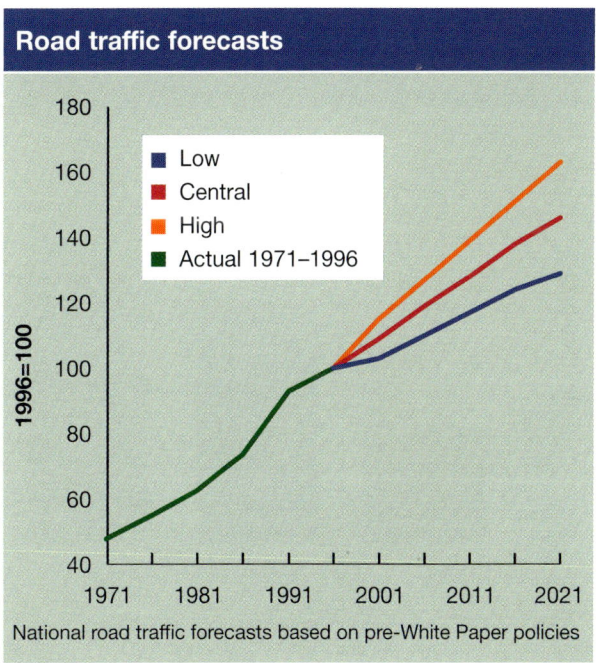

National road traffic forecasts based on pre-White Paper policies

1.11 We all know that unless something is done this means more traffic jams, not just in the cities but in country towns too. The tranquility of the countryside will be further eroded. Rush 'hours' will become longer. Driving will become even less of a pleasure and the costs to business will soar. There will be more damage to the environment and our health will suffer.

---

2   "Traffic in towns. A study of the long term problems of traffic in urban areas", HMSO 1963.

3   £15 billion taken from "Moving forward – a business strategy for transport", CBI 1995. Other estimates include £7 billion from National Economic Research Associates, July 1997.

CHAPTER 1   A New Deal for Transport

There is less choice ....

1.12 Increasingly, people do not have real choices. For many people using a car is now no longer a choice but a necessity. Nowhere is this clearer than in the rural communities with no daily bus service. For those who rely on public transport it is all too often inadequate, suffering from declining standards and services. And as motoring costs fell in real terms, bus and rail fares have gone up.

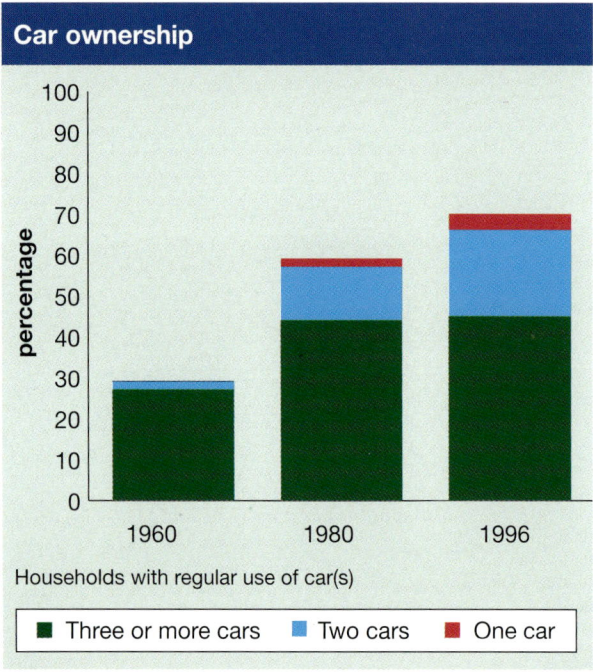

Car ownership

Households with regular use of car(s)

■ Three or more cars   ■ Two cars   ■ One car

1.13 Three in ten homes in Britain don't have a car – some thirteen million people. The advantages of owning a car aren't available to them. Even in homes with a car it is not always available to everyone.

1.14 Increased traffic, and speed, have made our streets more threatening for pedestrians and cyclists. Children's freedom to play, or to walk or cycle to school unaccompanied has been severely curtailed. Twenty years ago, nearly one in three 5-10 year-olds made their own way to school. Now only one child in nine does. We walk less than we used to and cycling, other than for leisure, is mostly left to a few enthusiasts.

1.15 In Britain, we have fewer cars but our cars do more mileage and we use public transport less than in most other countries in the European Union. It is not surprising that our roads are among the most congested. But it doesn't have to be like this.

.... and people want change

1.16 People know we cannot build our way out of congestion with new roads. The previous Government, too, came to see the problems – the growth in road traffic was at the heart of its national debate on transport. Its subsequent Green Paper "Transport: the Way Forward"[4] highlighted the dilemma of road transport – on the one hand the advantages it can bring but, on the other, the environmental damage it causes. Change was proposed – a new approach to transport policy that was not led by road building.

1.17 People feel the time for action is long overdue. The results of our consultation[5] last year confirm the overwhelming desire for Government to show leadership. People said they want more choice on whether to use their cars and more reliable journeys when they do; they want a better public transport system and one that doesn't let them down; they want better protection for the environment and they want less pollution because they are worried about their health.

---

4   and "Keeping Scotland Moving: A Scottish Transport Green Paper", Cm 3565, 1997. A summary of the main proposals in "Transport: the Way Forward" is provided in Annex D.

5   "Developing an Integrated Transport Policy. An invitation to contribute", Department of the Environment, Transport and the Regions, Department of the Environment for Northern Ireland, The Scottish Office and The Welsh Office, 1997. A summary of responses is provided in Annex B, with a fuller breakdown in the report on the consultation that we are publishing separately.

A New Deal for Transport

Pollution damages health.

1.18 Early in this Parliament we received a comprehensive report on transport and the environment from the Royal Commission on Environmental Pollution[6]. The Commission, too, raised concerns that "action has been too little and too slow" and warned that continuing as previously would have consequences that were environmentally, economically and socially unacceptable.

1.19 We agree. That is why we acted quickly and announced a fundamental review of transport policy. This White Paper is the culmination of that work and the widespread consultation that accompanied it.

A new approach: *A New Deal for transport*

1.20 We face an enormous challenge to deliver our vision of a transport system that supports sustainable development. We need a new approach, bringing together the public and private sectors in a partnership which benefits everyone. We want to ensure that companies have incentives to provide new services and raise standards, that taxpayers' money is spent wisely to make public transport available for all and that services are properly regulated in the public interest.

1.21 We have not put everything on hold until this White Paper. We are already working to extend the range of transport choices across the country and are investing more in public transport to improve its quantity and quality. We have secured new and imaginative ways of funding to modernise our transport system. We are giving high priority to maintaining and managing the nation's transport infrastructure. Taken together, this public and private investment represents a substantial increase in resources for transport. More investment in public transport and more people using it will work together to create a virtuous circle, generating more revenues, further investment and even better services.

1.22 We want transport to contribute to our quality of life not detract from it. The way forward is through an **integrated transport policy**. By this we mean:

- integration **within and between different types of transport** – so that each contributes its full potential and people can move easily between them;

- integration **with the environment** – so that our transport choices support a better environment;

- integration **with land use planning** – at national, regional and local level, so that transport and planning work together to support more sustainable travel choices and reduce the need to travel;

- integration **with our policies for education, health and wealth creation** – so that transport helps to make a fairer, more inclusive society.

1.23 **This is our *New Deal for transport* ....**

---

[6] "Transport and the Environment – Developments since 1994", Twentieth Report of the Royal Commission on Environmental Pollution, 1997. Cm 3752. ISBN 0-10-137522-0. A summary of the Report's main conclusions is provided in Annex C.

.... better places to live

1.24 We want a transport system that meets the needs of people and business at an affordable cost and produces better places in which to live and work. We want to cut congestion, improve our towns and cities and encourage vitality and diversity locally; helping to reduce the need to travel and avoid the urban sprawl that has lengthened journeys and consumed precious countryside. We will revise the planning guidance we issue to bring together thinking about better transport and a better environment at the planning stage.

1.25 The *New Deal for transport* means:

- cleaner air to breathe by tackling traffic fumes;
- thriving town centres by cutting the stranglehold of traffic;
- quality places to live where people are the priority;
- increasing prosperity backed by a modern transport system;
- reduced rural isolation by connecting people with services and increasing mobility;
- easier and safer to walk and cycle;
- revitalised towns and cities through better town planning.

.... local transport plans

1.26 We want to see integrated transport locally as well as nationally, which is why we are introducing *local transport plans* as a core part of our proposals. Local authorities will set out in these plans their strategies for transport. There will be new tools to tackle congestion and pollution which will provide local authorities with new and dedicated sources of funding for transport measures. But we are not relinquishing responsibility for what happens locally, we will need to be satisfied that these new powers will be used as part of clear transport strategies that have the backing of local communities.

1.27 We want more priority for public transport, improved facilities for people to interchange when travelling and better information for passengers. We want bus lanes that are properly enforced so that buses are more reliable as well as more frequent. We will improve choice and reliability of journeys in ways which safeguard the environment and the health of the nation. We will change the focus of road investment to reflect the needs of all road users, giving top priority to maintaining and managing our existing roads and getting them to work better.

1.28 We will put greater emphasis on listening to transport users – there will be a New Deal for the public transport passenger and a New Deal for the motorist. We will continue to work with motoring organisations to improve the service offered by trunk roads and, through investing in technology, we will improve the speed and efficiency of customer services provided by Government agencies.

1.29 The *New Deal for transport* means:

- new *local transport plans*:
  - integrated transport strategies for local needs;
  - local targets eg for improving air quality, road safety, public transport and road traffic reduction;
  - more certainty of funding;
  - greater use of traffic management;
- new powers including road user charging and levies on parking to tackle traffic jams and traffic growth;
- new sources of additional funding for local transport: better for the environment and better for business;
- better interchanges;

- tackling the 'pinch-points' in transport networks that lead to congestion;
- new airports policy and stronger role for regional airports;
- new independent *Commission for Integrated Transport* (CfIT) to advise on integration at the national level and act as a force for change.

A New Deal for the motorist

- improved management of the trunk road network to reduce delays, through eg Regional Traffic Control Centres in England
- investment focused on improving reliability of journeys
- better maintained roads – increased resources both locally and nationally
- updated Highways Agency's Road User's Charter to give more emphasis to customer service
- more help for the motorist if their car breaks down on a motorway
- reducing the disruption caused by utilities' street works
- improved road safety and safer cars
- quality information for the driver – before and during journeys
- dealing with car crime
- more secure car parks
- better information and protection when buying a used car
- action on 'cowboy' wheelclampers
- more fuel-efficient cars
- less congestion on our roads and less pollution in our cars

.... better buses

1.30 Buses will be cleaner, more comfortable and more reliable, a real and attractive alternative to using cars. We are challenging the industry to produce a bus design fit for the next century. We will build on *Quality Partnerships*, local partnerships to deliver better bus services. We will ensure that the passenger gets a real say in influencing bus services in their local area. *Quality Contracts*, where there is local demand, will mark a real change from the present and provide the opportunity for the development of integrated networks.

1.31 The *New Deal for transport* means:

- buses to lead our transport revolution for the 21st Century;
- upgraded *Quality Partnerships* between local authorities and bus operators:
  - quicker, more reliable services;
  - higher quality vehicles with staff trained in customer care;
  - easy-to-use buses – to help access for disabled and elderly people and parents with young children;
- *Quality Contracts* – exclusive contracts for bus routes to ensure integrated networks;
- half-price or lower concessionary fares for elderly people;
- special funding for buses in the countryside.

..... better trains

1.32 Through a new *Strategic Rail Authority*, we will bring vision to the privatised railway and we will ensure that it meets the needs of passengers and the freight customers it serves. Passengers rightly demand better services and more accountability. We are willing to re-negotiate existing rail franchises where this would secure benefits for passengers and value for money for the taxpayer.

CHAPTER 1   A New Deal for Transport

1.33 The *New Deal for transport* means:

- a new *Strategic Rail Authority* to:
    - bring together passenger and freight interests;
    - promote better integration and interchange;
    - provide strategic vision;
    - get better value for public subsidy in terms of fares and network benefits;
- new passenger dividends from passenger railway companies;
- tougher regulation to serve the public interest:
    - ensuring that the private sector honours its commitments to deliver a modern and efficient railway.

A New Deal for the public transport passenger

- more and better buses and trains, with staff trained in customer care
- a stronger voice for the passenger
- better information, before and when travelling; including a national public transport information system by 2000
- better interchanges and better connections
- enhanced networks with simplified fares and better marketing, including more through-ticketing and travelcards
- more reliable buses through priority measures and reduced congestion
- cash boost for rural transport
- half price or lower fares for elderly people on buses
- improved personal security when travelling
- easy-access public transport – helping disabled and elderly people, and making it easier for everyone to use

.... better protection for the environment

1.34 We want to preserve and enhance our environment: the places where we live and work, our built and natural heritage and our richly diverse countryside. We will be more effective in our stewardship of natural resources and are determined to build from the historic turning point of the special United Nations' conference at Kyoto, where the developed countries agreed to legally binding targets to reduce greenhouse gas emissions. We have already made an important step forward under our Presidency of the European Union (EU), reaching agreement on how to share the EU's target between Member States.

1.35 We want to see greener, cleaner vehicles that have less impact on our environment. We want to see better public transport and we will make it easier to walk and cycle. But these alone will not be sufficient to tackle the congestion and pollution that is caused by road traffic: we need to reduce the rate of road traffic growth. We also want to see an absolute reduction in traffic in those places and streets where its environmental damage is worst.

1.36 The *New Deal for transport* means:

- a major effort to reduce greenhouse gases;
- greener, more fuel efficient vehicles through:
    - better standards and tax incentives;
    - Cleaner Vehicles Task Force;
- better stewardship of the nation's cultural and environmental heritage;
- tackling transport noise and new powers to enforce noise controls at airports.

.... better safety and personal security

1.37 We want people to be able to travel safely and without fear for their personal security. Pedestrians and cyclists should not be intimidated by traffic; parents should not have to drive their children to school because they worry about their safety; women and older people should feel safe to use public transport after dusk.

1.38 The *New Deal for transport* means:

- root and branch review of transport safety;
- new road safety strategy and targets to reduce accidents;
- safer routes to schools;
- major review of speed policy;
- safer public transport;
- changes in drivers' hours legislation;
- review of the role and function of the British Transport Police;
- Secure Stations Scheme.

.... a more inclusive society

1.39 Because access to transport can be a matter of social justice we want to see high quality public transport designed for everyone to use easily. We want to tackle the downward spiral of disadvantage in the most deprived areas in the country, where difficulties in getting to jobs combine with other social and economic problems. Better transport is an essential building block of our New Deal for Communities which will extend economic opportunity, tackle social exclusion and improve neighbourhood management and quality of life in some of the most rundown neighbourhoods in the country.

1.40 As well as prosperous towns and cities we want a thriving countryside in which there are real jobs and opportunities for the people who live there. So where there is new development it should be planned in a way which supports existing communities. We know that transport needs vary widely within and between rural areas. The problems of remote island communities in Scotland are very different from rural villages in the South East of England. So will be the solutions.

1.41 The *New Deal for transport* means:

- more local diversity and vitality through better planning;
- opening up job opportunities:
  - through transport supporting regeneration;
- more and better buses;
- tackling isolation in the countryside through:
  - support for local facilities;
  - special funding for buses;
  - support for community projects to improve accessibility;
- tackling the transport needs of women, disabled and elderly people and people on low incomes;
- reuniting communities cut in half by traffic:
  - through traffic management, calming and traffic reduction;
- monitoring the impacts of policies on different groups in society.

.... moving goods sustainably

1.42 We are building a new partnership with business to improve the competitiveness of industry for the 21st Century. We want a reliable and efficient transport system that supports prosperity, to provide the jobs and wealth we all want. But the growth in freight risks being met at the expense of our environment. This is why we want to reduce the extent to which a healthier economy results in high levels of road traffic growth. We want to see a real increase in the use of rail freight, inland waterways and coastal shipping.

1.43 The *New Deal for transport* means:

- a new *Strategic Rail Authority* to promote rail freight and its infrastructure;
- *Quality Partnerships for freight* between local authorities and operators on lorry routing and delivery hours;
- less damage to roads and the environment through greater use of 6 axle lorries and keeping unsuitable lorries off unsuitable roads;
- working in partnership with the freight industry to improve best practice;
- impounding illegally operated lorries;
- facilitating shipping as an efficient and environmentally friendly means of carrying our trade;
- extending freight grants to include coastal and short sea shipping.

...... sharing decisions and modernising local democracy

1.44 We have made good progress in meeting the demand for decentralisation of power through our proposals for devolution. Different parts of the UK will be able to consider their own transport priorities reflecting their different transport needs. We also want to revitalise local democracy and strengthen the relationship between local and central Government. We will bring power closer to people and play our part in building effective partnerships.

1.45 We want local people and business to have a real say and real influence over transport. We will modernise the way in which transport is planned regionally and locally. We will expect local authorities when preparing their *local transport plans* to consult widely and involve their communities and transport operators in setting priorities for improving transport. In approving *local transport plans*, we will want to be sure that they fully reflect this consultation and that the views of local people have made a difference.

1.46 The *New Deal for transport* means:

- many decisions on transport issues to be devolved to the Scottish Parliament, the Welsh Assembly and the Assembly for Northern Ireland;
- strengthened planning arrangements in English regions to secure integration between transport and land use planning – including the role of airports, ports, railways and roads in the regions;
- Mayor for London and the Greater London Authority to produce an integrated transport strategy, improve air quality and act on noise;
- decision-making on transport to be more accountable to local people.

.... everyone doing their bit

1.47 Our *New Deal for transport* sets the framework for change and we will provide the new powers and extra support needed to make it happen. But we cannot do it alone. We want to create partnerships at all levels, to help business, local authorities and local communities to come together and respond to the challenge.

1.48 Much will depend on each one of us as individuals. For example, a significant reduction in car commuting and the 'school run' would help to tackle peak-time congestion. We cannot leave it to others to bring about the changes that are needed. We have a shared responsibility. But great sacrifices aren't called for. **It doesn't take much to make a difference – if we all left the car at home just once** out of the ten or so shopping and leisure trips we make from home each month, **we would deal with most of the projected increase in traffic this year**[7].

1.49 The *New Deal for tranport* means:

- Government departments taking the lead in introducing 'green transport plans' – plans which help to cut down on car use;

- local authorities, business, community organisations, schools and hospitals encouraged to produce their own green transport plans;

- a major national awareness campaign;

- new initiatives on school journeys;

- individuals/families/communities considering their own travel habits.

.... delivering the New Deal for transport

1.50 In this White Paper we set out the *New Deal for transport*. In Part II we look more closely at the problems that we have inherited and at why it is so important to set the right framework for change and have clear objectives. We describe the difference that our policies will make. We commit ourselves to challenging targets and rigorous monitoring and set out in detail the measures needed to secure changes on the ground.

1.51 In Part III we explain how the *New Deal for transport* will be supported by a new framework for action at national, regional and local levels and by getting the right balance between incentives, voluntary initiatives, best practice and economic instruments. We consider how we can all do our bit to produce a difference, explaining how the *New Deal for transport* supports and encourages local and individual action.

---

7   calculated from National Travel Survey 1994/96 data for home based journeys for the purpose of shopping, leisure, personal business (eg trips to the bank/hairdresser) and to see friends and relatives somewhere other than where they live

PART 2

CHAPTER 2   Sustainable Transport
CHAPTER 3   Integrated Transport

A New Deal for transport: better for everyone

CHAPTER 2
Sustainable Transport

Sustainable development meets the needs of the present without compromising the ability of future generations to meet their own needs.

From the Brundtland Report, 1987[1]

2.1 A modern transport system is vital to our country's future. We need a transport system which supports our policies for more jobs and a strong economy, which helps increase prosperity and tackles social exclusion. We also need a transport system which doesn't damage our health and provides a better quality of life now – for everyone – without passing onto future generations a poorer world. This is what we mean by sustainable transport and why we need a *New Deal*.

The New Deal for transport

2.2 We don't underestimate the difficulties. There is much that needs to be done to recover from the legacy we inherited. The lack of a strategic and integrated approach in recent years has made many of the problems worse. But our *New Deal for transport* sets out a framework for change.

2.3 It will be supported by clear and challenging targets, setting out what we want to achieve and by when. By publishing indicators we will be able to measure progress in a way that is clear and comprehensive. It will enable us all to see what is working and what more needs to be done.

2.4 It is a long term strategy to deliver sustainable transport. It is also a strategy for modernisation that harnesses the latest developments in technology. It begins in this Parliament, looks towards the next and sets out a programme for improving our quality of life for years to come. But to meet the country's needs, it must and will make a difference now as well as in the future. This Chapter sets out the framework for change and explains what that difference could be.

Better health

2.5 The way we travel is making us a less healthy nation.

Cycling: better for health, better for the environment.

2.6 Coronary heart disease is the biggest killer of adults in this country. Part of the blame is that we drive too much when we could walk or cycle. More exercise would help to reach the proposed target for reducing coronary heart disease and strokes in England, set out in "Our Healthier Nation"[2].

---

1   the most commonly used working definition of sustainable development – taken from "Our Common Future", (The Brundtland Report) – Report of the World Commission on Environment and Development, Oxford University Press, 1987 ISBN 0-19-282080-x.

2   "Our Healthier Nation. A Contract for Health", Cm 3852, 1998. ISBN 0-10-138552-6. "Working together for a healthier Scotland", Cm 3584, 1998. ISBN 0-10-138542-0. A Welsh Green Paper will be issued shortly.

2.7 Road traffic is a major contributor to air pollution. Up to 24,000 vulnerable people are estimated to die prematurely each year, and similar numbers are admitted to hospital, because of exposure to air pollution, much of which is due to road traffic[3]. Tighter standards and advances in vehicle design have helped to reduce those emissions which cause the greatest concern but in the longer term these gains could be at risk if traffic growth continues unchecked. Even this downward trend in emissions will not be sufficient in all places to reach our local air quality objectives set for 2005[4]. We must do everything we can to cut this loss of life by improving air quality, including further controls on vehicle emissions which have brought about significant reductions in emissions without imposing unreasonable burdens on car users or on business.

Reducing emissions to improve the air we breathe.

2.8 Motorists themselves and their passengers are at most risk from exhaust fumes. Recent studies[5] have shown that cars offer little or no protection against the pollutants generated by traffic. Car drivers face pollution levels inside a car two to three times higher than those experienced by pedestrians. Car commuters may receive more than a fifth of their total exposure to some pollutants from their daily journey to and from work, as well as adding to the pollution on our streets.

2.9 Although serious road casualties have declined, too many people are still killed or seriously injured on our roads (more than 120 people every day in 1997) and in other transport accidents. Some in society are more at risk. Children are particularly vulnerable and those from disadvantaged backgrounds are more likely to die as a result of road accidents than children from more affluent homes[6].

2.10 But the threat to children's health from the way we travel goes beyond accidents and pollution. Because of worries about safety, many parents now shuttle children to school by car when previously they would have made their own way on foot or by bike. The British Medical Association has warned[7] that the effects on children's physical health and mental development could be serious.

2.11 Traffic contributes substantially to the noise that has become part of the everyday environment and can make many people's lives a misery. There is now some evidence[8] that this noise disturbs sleep and affects performance in school children and that the stress this noise causes may increase the risk of developing chronic heart disease and psychiatric disorders. Noise is an important issue for those living close to airports and under flight paths and near to busy roads.

---

3   "Quantification of the Effects of Air Pollution on Health in the UK", Committee on the Medical Effects of Air Pollutants, Department of Health, 1998.

4   in the National Air Quality Strategy.

5   "Road user exposure to air pollution", a literature review published on behalf of DETR by Environmental Transport Association, 1997. ISBN 1-873906-14-5.

6   I Roberts and C Power, BMJ volume 313, 1996.

7   "Road transport and health", British Medical Association, 1997.

8   "The Non-Auditory Effects of Noise", Institute for Environment and Health, 1998.

2.12 The *New Deal for transport* therefore sets the framework to:

- reduce pollution from transport;
- improve air quality;
- encourage healthy lifestyles by reducing reliance on cars, and making it easier to walk and cycle more;
- reduce noise and vibration from transport;
- improve transport safety for users, those who work in the industry and the general public.

More jobs and a strong economy

2.13 The transport system moves goods and people and helps to make the economy tick. Good transport is needed to get people to work and many jobs are based on extensive travel. Transport is also a major contributor to the economy in its own right, currently employing around 1.7 million people[9].

2.14 We rely on efficient transport to ensure that goods and services are distributed throughout the UK and exported overseas. Yet in recent years investment in transport has failed to maintain the physical quality of the system, allowing valuable assets to deteriorate. There is a backlog of neglect of railway stations, track and bridges[10]; and roads in England and Wales are in their worst state for twenty years[11].

2.15 More than four-fifths of domestic freight tonnage goes by road. But traffic congestion now costs the nation billions of pounds each year and with traffic forecasts pointing to more congestion these costs can only increase. Important parts of our motorways suffer daily from traffic jams but building more roads can just encourage more traffic.

2.16 Modern business practices put firms at even greater risk from delay and congestion. 'Just in time' production, for example, means that companies no longer hold large stocks of raw materials, components or finished products on site, depending instead on their suppliers meeting their needs at short notice. They rely heavily on an efficient road network.

2.17 On the busiest roads in our cities journey times in the rush hour could lengthen dramatically, by as much as 70% over the next 20 years. Already in outer London one-fifth of the time taken to make a journey during rush hours is spent stationary. In central London, at any time of the day, drivers face the prospect of spending a third of their journey at a standstill[12]. Even our country towns at the busiest times can grind to a halt through congestion.

2.18 Rail freight tonnage has dropped by more than a quarter over the last decade, although the tide has turned in recent years. The lack of investment in rail infrastructure has led to increased delays and unreliability.

2.19 Air transport has been growing dramatically. But we haven't made the best use of the airports in our regions and we need to improve public transport to all our airports. Shipping is one of the most environmentally sustainable means of transport, carrying 95% of our growing

---

9   Office for National Statistics Labour Market Statistical Group.

10 see Railtrack's "Network Management Statement", 1998.

11 from the visual survey of the "National Road Maintenance Condition Survey 1997": the condition of roads in England and Wales was the worst recorded since the survey began in 1977.

12 time spent at 0 mph in a queue of traffic or spent waiting at traffic lights or road junctions. Taken from "Traffic Speeds in Central and Outer London: 1996-97", DETR, Statistics Bulletin (98) 17.

international trade by tonnage. The UK is a world centre of excellence for shipping and maritime-related activities. But recent decades have seen a massive decline in the size of our merchant fleet.

2.20 The *New Deal for transport* therefore sets the framework to:

- improve reliability for journeys in all modes, helping to support business and economic growth;

- improve links with international markets;

- support regeneration and the vitality of urban and rural areas;

- make more efficient use of the transport system;

- promote more sustainable UK transport industries.

A better environment

2.21 The way we travel is changing our environment for the worse. The 'skyglow' from light pollution and noise from transport have changed much of our countryside. Road construction and car parking have made heavy demands on land, a finite resource. In England alone, in the second half of the 1980s an area equivalent to the size of Bristol was taken for road building and parking[13].

2.22 Climate change is one of the greatest environmental threats facing the world today. Globally, the balance of evidence now points to a discernible human influence on the earth's climate through the emission of greenhouse gases. In the UK, transport's share of carbon dioxide (CO_2) emissions, the main greenhouse gas, has grown from around one tonne in eight in 1970 to more than one tonne in four in 1995, and is set to grow still further. Four-fifths are produced by road vehicles.

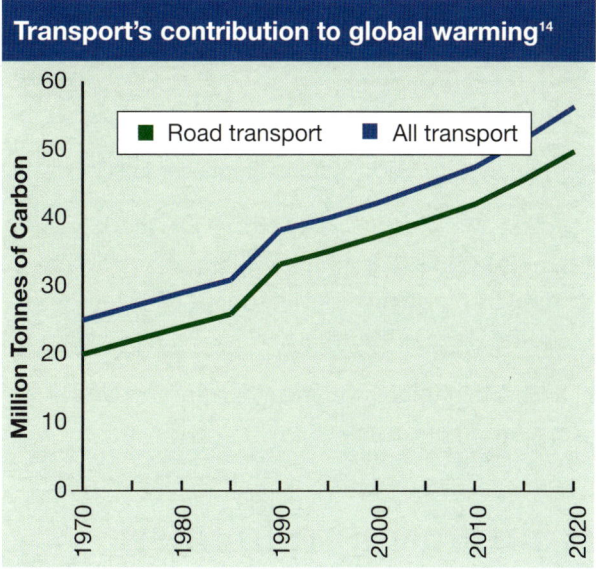

Transport's contribution to global warming[14]

2.23 As we use cars more, we have made less use of public transport. Yet buses and trains can have distinct environmental advantages as highlighted by the Royal Commission on Environmental Pollution. Buses require less road space per seat than cars and usually emit less CO_2 per occupant. Emissions of CO_2 and most other pollutants are lower per tonne-kilometre for rail freight than road freight. And emissions of CO_2 and most other pollutants are generally lower per passenger-kilometre for rail than for road.

2.24 We all know about noise pollution and road congestion around airports. But air traffic also has a global impact. CO_2 emissions per passenger-kilometre are higher from air travel than from most other ways of travelling and fuel for air travel now accounts for one-sixth of transport fuel sold in the UK.

2.25 The *New Deal for transport* therefore sets the framework to:

- reduce road traffic growth;

- respond to the challenge of climate change;

---

13   some 10,500 hectares in England changed to highways and road transport uses (public car parks and bus stations). Taken from Department of the Environment, Transport and the Regions' Land Use Change Statistics.

14   Transport CO_2 emissions. Source DTI EP65 central forecasts which are currently being revised for publication later this year.

- minimise transport's demand for land, protect habitats and maintain the variety of wildlife;

- limit the visual intrusion caused by transport;

- reduce use of non-renewable materials/energy sources;

- ensure that environmental impacts are taken fully into account in investment decisions and in the price of transport;

- enhance public awareness of transport and environment issues.

A fairer, more inclusive society

2.26 Nearly a third of households in Britain don't have a car – some 13 million people. The number who rely on public transport, walking or cycling is even higher because in those homes where there is a car not everyone has regular access to it. Those who can't drive have to rely on lifts (over 4 in 10 women don't have driving licences) and in many families there is a main driver who has 'first call' on the car. In some places, poor public transport and lack of a car combine to produce social exclusion. For example, some families in rural areas have had to make great financial sacrifices to keep a car to avoid relying totally on the little public transport that exists.

2.27 Most users of public transport rely on buses to get about. The less affluent – students, retired (there are five million elderly people without a car) and unemployed people – use buses more than others. It is these people who have had to face bus fares rising by almost a third in real terms since 1980. At the same time, the standard of living of bus and coach drivers has fallen – on average by 4% since 1985, compared with a 20% increase in real terms in the average wage.

2.28 Being unable to afford transport can limit everyday life. Job, training and education opportunities are more limited and there is less choice in shopping, adding to the family budgets of those least able to bear the cost. An expanded road network has helped people travel further and faster than before. But it has also led to jobs, shops and essential facilities moving out-of-town, reducing

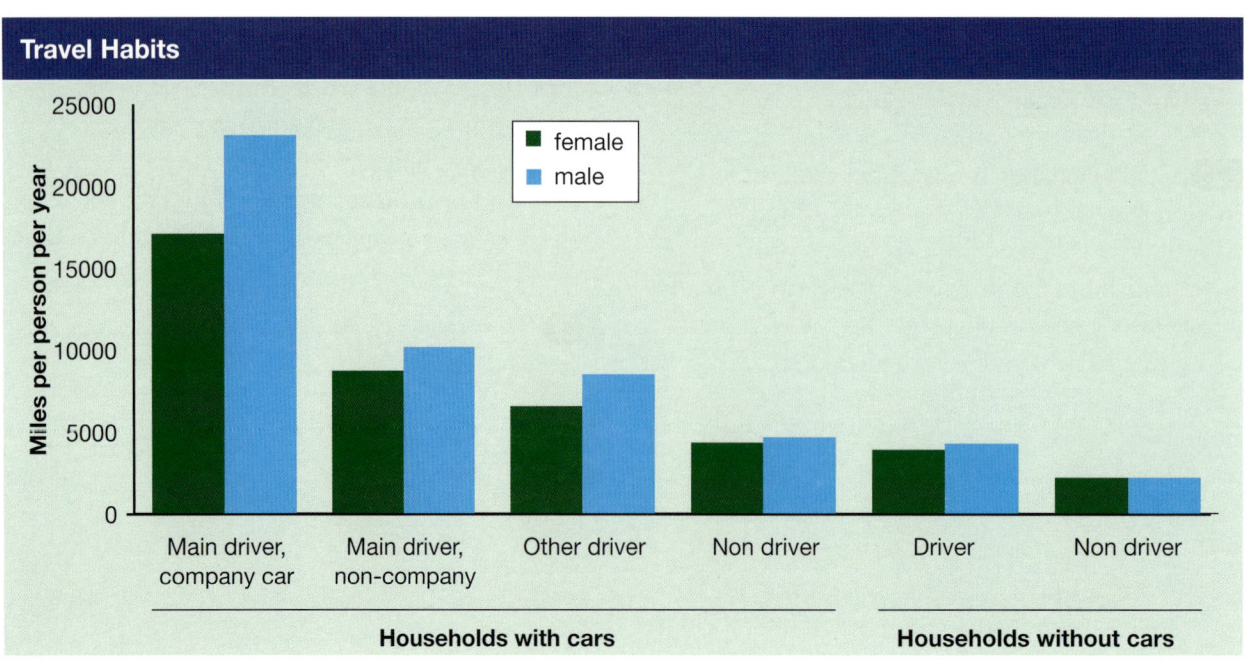

the vitality and diversity of local facilities and hitting the less mobile and those on low incomes.

2.29 Road traffic has affected some people more than others, the poorest and most vulnerable in society often suffer more than most. Busy roads in towns have cut communities in half and heavy traffic can be a barrier to community life. Road noise contributes to stress and disturbs sleep: those living closest to busy roads bear the brunt. Some of our town centres have been ruined by major roads, putting people in second place to the car. Increased traffic and speed have spoilt streets. Fear of traffic adds to the isolation sometimes faced by older people.

2.30 Public transport is not available to everyone, and where it exists is not always accessible to disabled people. Although recorded crime levels on public transport are low, concern for personal security is a significant deterrent to travel, particularly for older people, women and ethnic minorities.

2.31 The *New Deal for transport* therefore sets the framework to:

- produce better public transport and easier access to workplaces and other everyday facilities for all, especially people on low incomes;

- reduce community severance caused by transport;

- reduce the need to travel through better planning and technology;

- promote better transport choice for disabled people;

- reduce the fear of, and level of, crime on the transport system;

- promote better conditions for those working in transport.

A modern, integrated transport system

2.32 Privatisation, deregulation and competition were key features of the last decade but they have failed to deliver an integrated transport system. This needs to change. We want to work in partnership with industry but the shift of Government's strategic responsibility on to the private sector went too far.

MANAGING COMPETITION AND REGULATING MONOPOLIES

2.33 The legacy we inherited ranges from the competitive market of the deregulated bus industry to inadequate regulation of monopoly supply in the provision of railway infrastructure.

2.34 Whilst competition can bring benefits to some customers as suppliers compete for market share, the wider public interest must always be taken into account. In transport the problems of noise, congestion and pollution associated with individual travel decisions are often ignored and there is concentration on profitable routes at the expense of integrated transport networks which extend choice and accessibility.

2.35 We will therefore:

- build a framework which retains competition in the market but provides for intervention where there is evidence that this is needed in the public interest. The ability of competition authorities to deal with anti-competitive agreements and abuses of dominant position will be substantially improved by the provisions of the Competition Bill. Where operators deliver efficient services in the public interest they and their employees can expect to share in the rewards of their success;

- make increasing use of economic instruments such as pricing and taxation to send clear signals about the wider social and environmental impacts of travel decisions;

- improve the planning framework in a way which recognises the interactions between transport modes, land use and economic development, and provides for a more stable, integrated and strategic background within which transport operators and others may make investment decisions.

BUS DEREGULATION

2.36 Deregulation of the local bus market, outside London, caused substantial upheaval because of 'bus wars' and confusion over changing service patterns. There have been some good examples of innovation but frequent changes to bus services, poor connections and the reluctance of some bus operators to participate in information schemes or through-ticketing undermined bus services. In this climate, it was not easy for buses to match the levels of comfort, reliability and access offered by the private car.

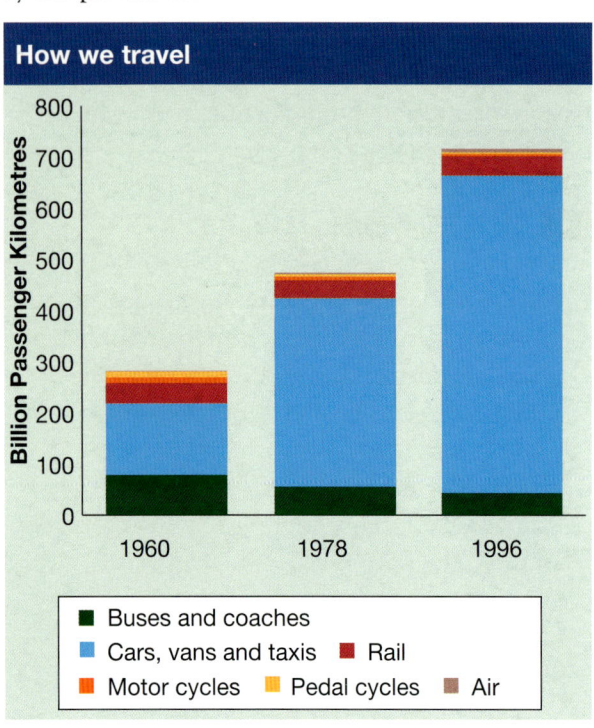

2.37 Deregulation has not broken the spiral of decline in local bus use. Since 1986 bus use has fallen by about a quarter – by about one billion fewer journeys a year; in contrast with London, within a regulated market, where use has held up. More recently, there have been good examples of bus companies and local authorities working together in Quality Partnerships to change the image of bus services and stem, sometimes even reverse, the decline in patronage.

RAIL PRIVATISATION

2.38 The previous administration supported the progressive liberalisation of access to the rail network by passenger train operators, who would compete with those already providing services. The Rail Regulator is legally committed to introduce greater competition from 1999. But open access with inadequate safeguarding of the public interest could lead to a loss of network benefits in areas like ticketing and timetabling. 'Cherry picking' of profitable routes could threaten local networks. This sort of behaviour has no place in our transport policy. The Rail Regulator has therefore set in hand arrangements to introduce limited competition subject to strict safeguards. Competition will not be allowed if it would undermine existing services supported by the taxpayer or reduce network-wide passenger benefits. The *Strategic Rail Authority* will be able to set the longer term policy framework for competition, ensuring continuing safeguards against erosion of a properly integrated network.

2.39 A healthy, growing economy has meant an increase in the number of rail passengers and this is welcome. But the privatised, fragmented railway that we have inherited is not making the most of this potential. And the privatised railway continues to receive vast amounts of public subsidy, with inadequate public accountability.

2.40 Some passenger train operators have gained new customers with better services and new products but the picture is patchy. For every train operator that has improved punctuality and reliability, there is another that has let standards slip: punctuality deteriorated in the year ending March 1998 in more than half the service groups operated and there were less reliable services in more than a third.

2.41 Passengers know that rail privatisation has not delivered the benefits claimed by its supporters. Figures compiled by the Central Rail Users' Consultative Committee indicate a substantial increase in the level of passenger dissatisfaction: in the first quarter of 1998 complaints almost doubled over the same period in 1997. The Committee had already expressed concern that there was "a gulf between what passengers can reasonably expect and what they receive and how it is delivered"[15].

2.42 The Rail Regulator published on 1 July 1998 figures showing that there were nearly one million complaints direct to train operators in 1997/8. That is a huge number. What is even more disturbing is the Rail Regulator's view that these complaints do not fully reflect passenger dissatisfaction with the privatised railway.

2.43 In a recent report, the House of Commons' Environment, Transport and Regional Affairs Select Committee highlighted the fundamental weaknesses in the organisation of the privatised railway[16]. These include the overlapping responsibilities of the Rail Regulator and the Franchising Director which leads to confusion about their respective roles; the inadequate sanctions when train operators and others perform badly; and the inconsistent regulation of key parts of the industry.

2.44 Critically, there is no good mechanism for long term strategic planning in the privatised rail industry. We will inject a strategic approach which will nurture the potential growth in rail patronage.

2.45 The *New Deal for transport* therefore sets the framework to:

- promote the public interest;
- produce better public transport, with easier, more reliable connections;
- improve choice between different modes;
- enhance public transport networks;
- encourage more through-ticketing;
- provide better travel information;
- ensure more reliable and frequent services;
- give the passenger a bigger voice in public transport.

Changing travel habits

2.46 Using the car less is not as impossible as some think it is. Nearly three-quarters of all journeys are under five miles and 45% are less than two miles. Even though many of us could walk or cycle these short distances, or catch a bus, we have increasingly used our cars – a quarter of all car journeys are now under two miles.

2.47 A recent study for the RAC[17] concluded that most car trips do not have to be made by car.

---

15 The Central Rail Users' Consultative Committee (CRUCC) is the statutory consumer organisation representing the interests of rail users nationally. Figures on complaints taken from CRUCC press release 9/98, dated 4 June 1998 and commentary from CRUCC press release 3/98, dated 16 March 1998.

16 Third Report of the Environment, Transport and Regional Affairs Committee, Session 1997-8, on the proposed *Strategic Rail Authority and Rail Regulation*, House of Commons paper 286-I, March 1998.

17 "Car dependence", a report for the RAC Foundation for Motoring and the Environment by ESRC Transport Studies Unit, University of Oxford with RDC Inc, San Francisco, 1995. ISBN 0 86211355 5.

Using a car currently seems the sensible choice because of factors such as physical and time constraints and the poor quality of alternatives. Some car trips (up to 30%) were judged to be hardly necessary at all or a perfectly good alternative was already available but ignored. This shows the potential for people to use their cars less without making great sacrifices – and often benefiting instead from the exercise, the stress avoided and the money saved.

2.48 We know that a very high proportion of people change their travel choices from day to day and year to year, showing a great adaptability in arranging their travel and their lives. The *New Deal for transport* will make it easier for people to choose different and more sustainable ways of making their journeys, helping them to make the changes in travel behaviour that are needed.

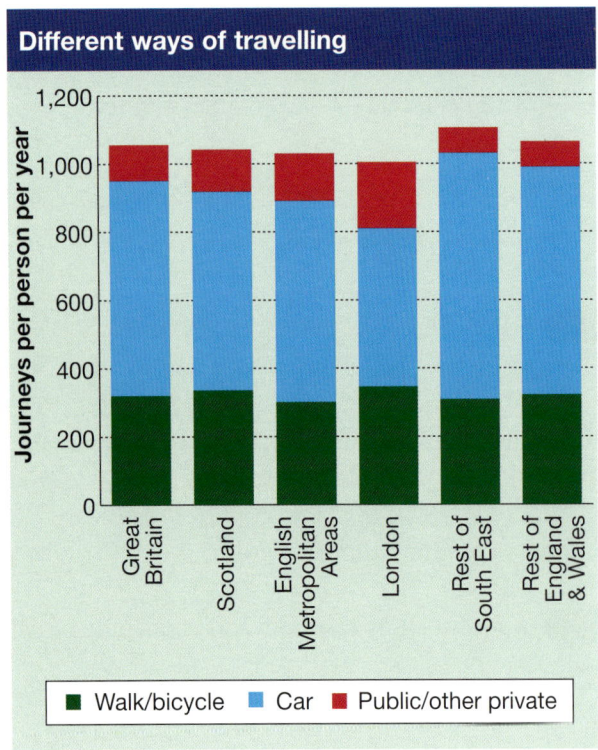

Technology taking the strain

2.49 We can all do our bit to make a difference and this will be helped by advances in technology. We are committed to making the fullest possible use of new technologies to deliver the *New Deal for transport*. As technology works best when combined with other measures, for example, financial incentives to buy greener vehicles, we will bring forward packages of measures to get the most from technological progress.

2.50 Improvements in fuel and vehicle technology, for example, will make a significant contribution to achieving our targets for improving air quality and reducing greenhouse gases. And developments in information technology will produce more reliable and comprehensive information to help public transport users and motorists plan their journeys. Technology can also help to make transport safer through, for example, improvements in vehicle design and the use of CCTV.

The New Deal for transport – making a difference

2.51 We have embarked on a comprehensive agenda for change – a series of practical, carefully thought out reforms. Our new approach will work best when the measures are combined in packages, so that each reinforces the other. We set out the measures in the remainder of this White Paper. Together, these measures will deliver the change that is needed. This integrated approach is vital if we are to meet the objectives and targets in our *New Deal for transport*.

2.52 We have international and national targets for protecting the environment. The Royal Commission on Environmental Pollution has produced two comprehensive reports on reducing transport's impact on our environment and proposed targets to drive the process. These have been key influences on our *New Deal for transport*. Challenging targets are helping to focus attention on reducing greenhouse gas emissions, improving local air quality and road safety, boosting rail freight and encouraging more cycling.

International and National Targets and Standards

- current targets:
  - greenhouse gases – legally binding target to reduce emissions to 12.5% below 1990 levels by the period 2008 to 2012 and a domestic aim to reduce CO_2 emissions by 20% by 2010;
  - air pollution – National Air Quality Strategy, encompasses health-based objectives for a range of air pollutants to be met by 2005;
  - EU vehicle and fuel quality standards – to reduce toxic emissions and noise from new vehicles;
  - cycling – from a 1996 base, double cycling by 2002 doubling again by 2012 (from the National Cycling Strategy);
  - road safety – existing target for 2000, new target for 2010.
- targets for the future:
  - freight on the railway – endorsement of the industry's targets for growth;
  - EU vehicle standards – target to improve fuel efficiency and reduce CO_2 emissions by more than a third before 2010;
  - health – proposed targets in "Our Healthier Nation" for reducing all accidents by a fifth by 2010 and reducing death rates from heart disease and strokes amongst people under 65 by a third by 2010;
  - green transport plans – for HQ/other key Government buildings by 1999/2000;
  - walking – targets being prepared to reverse the decline in walking;
  - public transport – targets to encourage more use of public transport;
  - road traffic – assess impact of measures in this White Paper and consider national targets for the level of road traffic.

Making a difference: on climate change

2.53 Following the Kyoto climate change conference in December 1997, the UK has a legally binding target to reduce greenhouse gas emissions to 12.5% below 1990 levels by the period 2008 to 2012. This means a reduction equivalent to 27 million tonnes of carbon. We also have a domestic aim to reduce CO_2 emissions in the UK to 20% below 1990 levels by 2010. We do not underestimate either the scale of the challenge or the huge potential for industry in the UK to benefit from improved fuel efficiency and win exports through developments in environmental technology.

2.54 We shall be consulting soon on options to meet our legally binding target and help move towards our domestic aim. The consultation will present an opportunity to assess savings from the transport sector in relation to measures which could be taken in other sectors. We will consider the balance of our programme in the light of public debate and responses during the consultation period.

Kate's story[18]	The New Deal for transport
My first job in the morning is to get the kids to school. By car, of course. It's too dangerous to let my nine year old, Sam, walk by himself – there is no lollipop lady where he needs to cross a busy road. Rebecca, who's 13, wants to cycle but I'm frightened that she'll have an accident.	We will work with local councils to make walking safer and to provide more cycle routes to schools. Schools will be encouraged to improve facilities for cyclists.
We need two cars. David, my husband, drives to work as well. He really has no other option. There's plenty of buses but they're dirty and unreliable and take longer than the car. But he does find the drive stressful. The traffic gets worse. He arrives at the office wound-up.	We will improve bus services. There will be investment in better information. New rules will improve the quality of the buses. And, by giving them priority in the rush 'hour', they will become more reliable.
On the way home, I sometimes pop down to the supermarket in the car. That allows me to get some shopping done without having to go out at the weekends. If I've time, I do it before picking up the kids. Trying to go round a supermarket with them is a nightmare.	Some supermarket chains are already introducing home shopping by phone, fax and the internet. This will be more convenient for many shoppers and reduce the number of car journeys.
We hardly ever use public transport. Or walk for that matter. David even drives the 500 yards to the paper shop on a Sunday morning.	We should all try to walk more – for our own health. Together with local councils we will help by making walking more pleasant, by cutting traffic and car speeds.

18 this and the other stories in this Chapter are composite pictures drawn from the widespread public consultation exercise and are designed to illustrate the concerns that people have about transport.

2.55 A range of measures will be needed from the transport sector. These will form an essential part of a balanced approach to reducing greenhouse gas emissions. Of the initiatives in the *New Deal for transport*, those directed at improving the fuel efficiency of all vehicles on our roads, especially those which target the fuel consumption of the cars we drive, have the greatest potential to reduce transport CO_2 emissions. Reducing the overall rate of road traffic growth and local action designed to achieve a switch to less fuel intensive forms of transport will also play an important part.

2.56 EU Member States have agreed a strategy for reducing CO_2 emissions from new cars, with the aim of achieving a reduction of more than a third by no later than 2010. This strategy alone has the potential to reduce forecast road transport CO_2 emissions in the UK by 8-14%[19]. But these improvements critically depend on the way we respond as car buyers. Motoring taxation also has an important role to play, by providing the right incentives for those decisions.

2.57 The main aim of our motoring tax strategy is to encourage people to buy more fuel efficient models and to invest in regular maintenance and fuel saving technologies. For example, recent analysis by the AA suggests that car owners can reduce fuel consumption and CO_2 emissions from their cars by as much as 5% by fitting low rolling resistance tyres[20].

2.58 Tackling congestion will also help improve the fuel efficiency of urban travel and reduce the impact of UK road transport on climate change. Even cautious estimates[21] suggest that fuel consumption, and therefore CO_2 emissions, are at least 10% higher on urban roads and 25% higher in the centres of our largest cities as a result of congestion.

2.59 The impact on emissions of the *New Deal for transport* depends on how quickly packages of measures to tackle congestion can be brought forward and implemented by local authorities. The potential gains are large. They can lead to absolute reductions in local traffic levels by promoting transfers to public transport and improvement in the fuel efficiency of the vehicles that remain. We estimate that getting the right transport packages in place locally, backed by road user charges, could deliver reductions of up to 20% in road traffic CO_2 emissions in the centres of our busiest cities[22].

2.60 The overall contribution of the measures set out in this White Paper towards reducing UK road traffic CO_2 emissions will depend on a great many factors, in particular the way in which local authorities, businesses and all of us as individuals respond to the lead it provides. Against this background, estimating a range for the potential savings is difficult.

2.61 We have based our analysis on the 1997 National Road Traffic Forecasts. Our work to date therefore includes the reduction of 14% in CO_2 emissions which is implicit in these forecasts, when compared with those published in 1995 for CO_2 emissions and shown in the earlier chart, 'transport's contribution to global warming'. What our work shows is that, even without a major change in behaviour, with the key measures in the *New Deal for transport*, there is the potential to reduce forecast 2010 road traffic CO_2 emissions by

---

19 compared to estimates based on the 1997 National Road Traffic Forecasts with unchanged traffic assumptions. The analysis assumes that the EU objective of reducing average (sales weighted) new car CO_2 emissions to 120 grammes per kilometre is achieved in the UK by 2005 or 2010 at the latest.

20 "Tracking CO_2 emissions from UK Home and Cars", Automobile Association, December 1997.

21 based on analysis using the framework developed for the 1997 National Road Traffic Forecasts and estimates produced by the Highways Agency of the impact of traffic speeds on CO_2 emissions.

22 from analysis based on the 1997 National Road Traffic Forecasts and published studies of local packages based on central area cordon charges and complementary public transport improvements for locations such as Bristol, Edinburgh and London.

Martin's story	The New Deal for transport
I drive to work. It's frustrating, though. The traffic is terrible and getting worse. I try to find new routes but there's no way to avoid it.	Better traffic and road management will help to cut congestion. The Highways Agency is developing a 'Toolkit' of ways to keep the traffic moving. We want to ensure there are decent alternatives to the car.
I do a lot of car journeys with my work, seeing clients and so on. I'm often late because of roadworks. And the state of the roads after they've finished is a disgrace.	We will give priority to maintaining the roads we have before we build new ones. And we are looking at ways to improve co-ordination between utility companies for essential streetworks.
You can suddenly hit really bad traffic with no chance to take a different route.	We are working with the private sector to improve information systems on traffic problems, both before you set off and while you are driving.

22-27%[23]. Other measures in the *New Deal* which are less easily quantified will add to this. With a step change in attitudes even greater reductions are within our grasp.

Making a difference: on traffic and congestion

2.62 Increased traffic and congestion is at the heart of many of the problems we face. Local traffic authorities are already required to consider setting targets for traffic reduction in accordance with local circumstances. The amount of traffic on our roads will be a powerful indicator of how things are going, at both national and local levels. We agree with the Royal Commission on Environmental Pollution that national targets for the reduction of road traffic must have a clear and specific justification in terms of the environmental benefits they are expected to achieve, and must represent the preferred and most effective method of achieving those objectives.

2.63 We will therefore assess the broad impact on national road traffic levels of the measures we are proposing and, in the light of that assessment, consider how national targets can best help. The question of national targets for road traffic reduction has been debated by Parliament in the context of the Road Traffic Reduction (National Targets) Bill. In considering national targets, we will seek advice from the *Commission for Integrated Transport*, the new independent body that we will set up to review progress on implementing our integrated transport policy (see Chapter 4).

2.64 We know that the *New Deal for transport* will make a difference at the local level. Analysis based on studies in cities such as Edinburgh, Bristol and London points to a reduction in public and private transport peak journey times of as much as 20-25% in the centres of the largest urban areas by 2010. This reduction in congestion would bring benefits to business and the environment. This underlines the conclusion of the Standing Advisory Committee on Trunk Road Assessment (SACTRA) in its interim report[24].

---

23 compared to the 1995 CO_2 emissions forecasts referred to above, and including the impact of the 1997 National Road Traffic Forecasts.

24 "Transport Investment, Transport Intensity and Economic Growth: interim report", SACTRA, 1997.

Making a difference: on local air quality

2.65 We estimate that European initiatives aimed at tightening vehicle and fuel standards have the potential to reduce busy central urban area road traffic nitrogen oxide emissions by up to 67% and particulate emissions by up to 70% below 1996 levels by 2010. Action at the local level, bringing together user charges and complementary public transport packages directed at tackling congestion and bus and freight *Quality Partnerships* directed at promoting cleaner buses and lorries[25], have the potential to deliver further significant savings. Further reductions in particulate emissions of up to a half appear possible.

2.66 Reducing urban road traffic emissions will make our towns and cities healthier places in which to live and work, bringing benefits in particular for those suffering from respiratory disorders including the increasing number of children with asthma. Better air quality will also benefit drivers and passengers who are currently exposed to high levels of pollution in busy city centres.

Making a difference: a more inclusive society

2.67 The *New Deal for transport* will produce a step change in public transport bringing significant benefits to both town and country with better mobility for all in society. The measures we are introducing will tackle the transport needs of women, disabled and elderly people and people on low incomes. Reducing road traffic in city centres will make it easier for local authorities to reallocate road space to cyclists and pedestrians.

Joe's story	The New Deal for transport
I'm a pensioner and I've never had a car. I've always used the bus. The service isn't too bad but it's pricey. Our local council doesn't have a pensioners' pass scheme.	The *New Deal* will mean a half fare pensioner's bus pass. That will be the minimum guarantee. Councils can have more generous schemes if they wish.
Once you get into town, the traffic is terrible. You have to be fairly quick to get across the crossings.	There will be more pedestrianised streets in town centres and more space for people who walk. We will encourage councils to design crossings so that people are not forced to deal with difficult subway crossings or bridges.
Sue's story	
I live in London and use public transport a lot. But I am worried about safety. You read about attacks and muggings on trains and stations. I don't like using them at night. I prefer to use the car because it's safer.	We will improve safety at stations and on public transport. We will encourage better lighting in stations and at station car parks. We will extend the use of CCTV. We will review security on trains with the operators.

---

25 for example by encouraging operators to fit particulate traps.

Making a difference: through extending the range of targets

2.68 We will help to draw up new targets – for example, for promoting public transport. In doing so, we will balance the costs and benefits of such targets and seek advice from the *Commission for Integrated Transport* on the form they might take.

2.69 At the same time, we will publish new indicators so as to allow progress to be measured. For example, in England a series of indicators is being developed by the Highways Agency to report on the performance of the trunk road network. We have already published our first report on *"Transport Trends"*, containing key indicators covering a wide range of transport topics: for example, on the use of different forms of transport, on transport safety, changes in the level of air pollutants, greenhouse gas emissions and road traffic.

2.70 We will continue to publish these reports each year so that progress against key indicators can be monitored. The indicators will include breakdowns by income groups, rural and urban areas and by age and gender so that we can monitor the impacts of policies on different groups in society. We will carry out the further work that is needed in some areas to ensure that targets and indicators are appropriate and effective. One such area is noise.

2.71 But in most cases we expect targets to be developed as part of coherent regional and local strategies to support integrated transport, rather than being set at the national level. National targets can usefully act as benchmarks and encourage improvement but they do not recognise local variations or draw on local knowledge of what is achievable. We suggest in Chapter 4 what these local targets could include. Drawing up targets regionally and locally will help to sharpen the focus of local policy; complementing the approach we have set for reaching our target in England to build 60% of new homes on previously developed land[26].

---

26 "Planning for the Communities of the Future", Cmd 3885, 1998. ISBN 0-10-138852-7.

CHAPTER 3
Integrated Transport

"Developing an integrated transport policy represents a major shift in direction. We don't just want to stop traffic problems getting worse, we actually want to make things better for people and goods on the move."

John Prescott
1997

More choice

Making it easier to walk

3.1 We are all pedestrians, even if we own a car. Nearly all journeys involve a walk and walking is still the main way of getting about locally. But all too often the things that make walking a more pleasant experience have not been given proper attention, as can be seen in the way road space and priority is so often biased against pedestrians. Too often pedestrians are treated like trespassers in their own towns. We want streets that are decent and attractive to walk in.

3.2 Too many of us have given up walking short distances in favour of using the car. We need to reverse that trend for the sake of our own and others' health, and for good environmental reasons.

3.3 Our *New Deal for transport* will make walking a more viable, attractive and safe option. Strategies to make it easier to get around locally by walking will be included in the *local transport plans* that we will introduce (see Chapter 4).

3.4 Reflecting our proposals for 'streets for people' that we describe later in this Chapter, we will expect local authorities to give more priority to walking by:

- reallocating road space to pedestrians, for example through wider pavements and pedestrianisation;
- providing more direct and convenient routes for walking;
- improving footpath maintenance and cleanliness;
- providing more pedestrian crossings, where pedestrians want to cross;
- reducing waiting times for pedestrians at traffic signals and giving them priority in the allocation of time at junctions where this supports more walking;

Making it easier to walk: an artist's impression of a pedestrianised Trafalgar Square.

- dealing with those characteristics of traffic that deter people from walking;
- introducing traffic calming measures near schools, in 'home zones' and in selected country lanes;
- using their planning powers to ensure that the land use mix, layout and design of development is safe, attractive and convenient for walking.

> **Better for pedestrians**
>
> - pedestrians in Birmingham's Bull Ring shopping centre will no longer get second class treatment;
> - the Bull Ring redevelopment gives better pedestrian links with the City's main shopping streets, has a new public square and gets rid of the warren of underground subways which people were forced to use previously.

3.5 We are working closely with local government and a wide range of organisations to prepare a strategy[1] that will provide a framework for action. In addition to supporting and developing this strategy, we will revise existing advice and work with local authorities and others in improving the environment for walking.

3.6 We will also encourage local authorities to introduce facilities which make it easier and safer for disabled and elderly people to move about. This will include pedestrian crossings that are fully accessible to all, including people in wheelchairs, and incorporate tactile features and audible signals to help blind and partially sighted people.

3.7 We have already made a start in promoting walking, and cycling, as healthy modes of transport through the 'Active for Life' physical activity campaign run by the Health Education Authority.

> **Safer routes to school**
>
> - the Myton cycleway makes it possible for the first time to cross the River Avon between Leamington and Warwick by cycle and on foot;
> - it links major housing areas on one side of the river with three large schools on the opposite bank;
> - benefits include less traffic at school times, reduced emissions and better longer term health of students and residents.

Making it easier to cycle

3.8 The National Cycling Strategy (NCS) published in 1996 highlighted the potential of cycling as a flexible, relatively cheap and environmentally friendly way to travel with important health benefits for people of all ages. We agree. Cycling, however, has been in decline nationally, even though more cycles are owned than ever (and annual sales of bicycles outstrip the number of new cars sold). But this doesn't have to be the case if we make it easier and safer to cycle:

- in Munich, cycle use rose from 6% of all trips in 1976 to 15% in 1992;
- in Hanover, cycling has increased from 9% in 1976 to 16% in 1990;
- in York in recent years about 20% of commuting has been by bike.

3.9 The NCS encourages local authorities and others to establish local targets for increased cycle use. A number have already done so and we expect targets to become more widespread as local strategies for cycling evolve. The NCS has established a national target of doubling the amount of cycling within six years (against a base year of 1996) and

---

1 being prepared by a steering group which comprises representatives of local and central government and a wide range of organisations and individuals.

of doubling it again by the year 2012. **We endorse this target.** A National Cycling Forum[2] has been established to oversee its implementation.

National Cycle Network

3.10 To support the NCS, we are continuing to research innovative measures to improve the safety and convenience of cycling and will publish advice on good practice. We want to see better provision for cyclists at their destinations, at interchanges, in the design of junctions and in the way road space is allocated. In particular, we are looking to local authorities to:

- establish a local strategy for cycling as part of their *local transport plans*;

- institute 'cycle reviews' of the road system and 'cycle audits' of proposed traffic schemes;

- adapt existing road space to provide more cycle facilities;

- make changes to traffic signalled junctions and roundabouts in favour of cyclists, giving them priority where this supports cycling;

- apply speed restraint more widely to support their cycling strategies and provide for cyclists when applying speed restraint measures;

- increase provision of secure parking for cycles;

- maintain cycle lanes adequately to avoid hazards to cyclists;

- use their planning powers to promote cycling through influencing the land use mix, layout and design of development and through the provision of cycle facilities.

3.11 Concern about road safety is a major reason for people not using their bikes for everyday journeys. Parents in particular see the dangers for their children of cycling on roads. In many areas radical changes are needed to create safer cycling conditions. Cycling promotion policies therefore need to mesh with those on road safety. Safety should be an additional incentive for action, not a reason for delaying priority measures for cyclists.

3.12 We will continue to help with the development of the National Cycle Network being co-ordinated by the transport charity Sustrans. The network will be a linked series of traffic-free paths and traffic-calmed roads providing some 8,000 miles of safe and attractive routes by 2005. By opening up opportunities for people to cycle more, the network will help to create a culture that welcomes cycling as an activity.

---

2   the Forum comprises representatives from a range of organisations across the UK including central and local government, business and the voluntary sector.

CHAPTER 3   Integrated Transport

Adapting existing road space for cyclists (Reading).

More and better buses

3.13 Buses are already the workhorses of the public transport system and in many parts of the country they are the only form of public transport. Increasingly they will become the focus of an efficient transport system that gets people to where they want to be quickly and comfortably, without having to rely on cars.

3.14 But people will not switch from the comfort of their cars to buses that are old, dirty, unreliable and slow. Too often buses have been treated and seen as 'second class' transport. It doesn't have to be like this and is certainly not the case in many other European countries.

3.15 As part of the *New Deal for transport* we want better buses – clean, comfortable and convenient. Bus lanes and other priority measures will help to get buses running on time. A first-rate and modern bus industry will make an important and cost-effective contribution to tackling congestion and pollution at the local level. By giving buses greater priority and improving information and networks, we can encourage more people to use buses. Increasing passenger numbers could transform the economics of bus operations, opening new horizons in quality, reliability and network expansion.

Quality Partnerships

Quality Partnerships have been developed in a number of towns and cities, eg in Aberdeen, Birmingham, Brighton, Edinburgh, Ipswich, Leeds and Swansea. They:

- include features such as new, higher quality and more accessible buses;

- have increased patronage by 10-20% and by up to 40% with bus segregation and substantial improvements in infrastructure;

- carry new passengers who previously used cars and taxis as well as those who walked.

Quality Partnerships can also be successful in rural areas, eg in Pwllheli, North Wales, where:

- a quality bus network has been maintained and patronage has remained buoyant thanks to Gwynedd Council working closely with bus operators, many of them small local firms;

- bus subsidy support has been combined with payments for school children's tickets (schools transport under the Education Act) and has prompted better services from operators including investment in new vehicles.

3.16 The most significant improvements in bus services recently have been achieved through co-operation between local authorities and operators under 'Quality Partnerships'. In these partnerships, the local authority provides traffic management schemes which assist bus services (bus lanes, priority at junctions, park and ride). The bus operator offers better quality (in terms of comfort, 'greenness,' accessibility and staff training), improved marketing, better integration and more reliable services.

3.17 Quality Partnerships work but they need to be more widespread and put on a firm footing. **We will therefore introduce legislation to put these partnerships on a statutory basis.** This will enable local authorities to require operators to meet certain standards of service quality in order to use the facilities provided by the local authority as part of the Quality Partnership. This will give local authorities greater influence over the provision of bus services and their marketing, and will enable them to encourage the provision of easy access buses.

3.18 Quality Partnerships should be for rural as well as urban areas, although a rural Quality Partnership might well look different. It might feature improved bus stops and information as well as higher quality vehicles, rather than traffic management. We have already made significant strides in improving bus services in the countryside and more Quality Partnerships will help. We set out our proposals for rural bus services in Chapter 4.

3.19 We will clarify local authorities' powers to buy in extra services to boost frequencies on a particular route or corridor. This will help to make bus use more attractive, particularly to those who would otherwise use cars.

3.20 In some circumstances, strengthened Quality Partnerships may not be sufficient to guarantee the necessary improvements. **We will therefore introduce primary legislation to give powers to local authorities, where it is in the public interest, to enter into *Quality Contracts* for bus services.** *Quality Contracts* would mark a real change from the present and would involve operators bidding for exclusive rights to run bus services on a route or group of routes, on the basis of a local authority service specification and performance targets. We will apply the experience from the best value approach to contracting which we are introducing to improve the quality and efficiency of services in local government. *Quality Contracts* will be subject to Ministerial consent for each local authority that wished to adopt such an approach (and following devolution, the consent of the National Assembly for Wales or the Scottish Executive).

Putting buses first: bus priority lane in Edinburgh.

3.21 The circumstances in which *Quality Contracts* might be considered will be the subject of national guidance, drawn up in consultation with local government. Initially, a small number of pilots could be used to demonstrate the contribution of *Quality Contracts* to developing bus networks and responding to what the passenger wants.

CHAPTER 3 — Integrated Transport

Putting buses first

The Leeds 'guided bus':

- quicker journeys in the morning peak;
- passengers think the service has improved;
- new and increased patronage.

'Greenways' in Edinburgh:

- in the first six months of operation average bus journey times cut by 25% on the all-day Leith Greenway and by 10% on the peak hour Corstorphine route;
- an additional 250,000 passengers travelled on Lothian Region Transport buses running on the Corstorphine and Leith Walk Greenways.

3.22 Listening to the passenger is an important part of the *New Deal for transport*. We therefore welcome the recent initiative by the Confederation of Passenger Transport to establish an independent Bus Appeals Body to handle bus passenger complaints outside London not resolved direct with operators. In London, this task falls to the London Regional Passengers' Committee.

3.23 We want a wider role for the Traffic Commissioners in strengthening the passenger voice. The Traffic Commissioners have an important independent oversight of the bus registration system and in licensing operators as fit and proper persons to operate bus services. We are considering with the Traffic Commissioners how best their role might be enhanced in delivering integrated transport.

3.24 Innovation is an important part of providing better bus services. For example, the use of smaller buses has become increasingly common. They can get to places where the traditional double-deckers would be inappropriate; they can provide more frequent services; and they can exploit niche markets that only require small buses. Taxis can also act as small buses although the use of the powers in the 1985 Transport Act has been disappointingly low. Local authorities will need to assess the potential for smaller buses and taxi buses, particularly in rural areas, when preparing their *local transport plans*.

Making a difference for the public transport passenger

- more and better buses and trains, with staff trained in customer care
- new *Strategic Rail Authority* to:
  - promote better integration and interchange
  - get better value for public subsidy in terms of fares and network benefits
- new passenger dividends from passenger railway companies, including more effective penalties to improve reliability and punctuality
- tougher rail regulation to serve the public interest:
  - ensuring that the private sector honours its commitments to deliver a modern and efficient railway
- a stronger voice for the passenger
- better information, before and when travelling; including a national public transport information system by 2000
- better interchanges and better connections
- enhanced networks with simplified fares and better marketing, including more through-ticketing and travelcards
- more reliable buses through priority measures and reduced congestion
- cash boost for rural transport
- half price fares or lower, for elderly people on buses
- improved personal security when travelling
- easy-access public transport – helping disabled and elderly people, and making it easier for everyone to use

3.25 We have also seen innovation in the structure of bus fares. For example, Magicbuses in Manchester have cheaper fares but are less luxurious than other buses on the same routes. Magicbus fares are typically 20% cheaper than alternative services. Results show that some passengers wait for a Magicbus, letting the better quality bus go. Others let the Magicbus go and prefer a better quality bus. Our proposals on fares are explained in Chapter 4.

A better railway

3.26 With the *New Deal for transport* there is the potential for a railway renaissance. But this will not be possible with the weaknesses arising from the fragmentation of the rail industry. **We will therefore establish a national *Strategic Rail Authority* for Great Britain**, to provide a clear, coherent and strategic programme for the development of our railways. This proposal is explained in Chapter 4, together with our new approach to franchising and investment in rail.

BETTER FOR PASSENGERS

3.27 Passenger rail services in Great Britain are provided by 25 franchised train operating companies, owned by 12 different franchisees, four of whom are also major operators of bus services. Recent performance of the privatised railway has not been good. But there is clearly scope for increased use of the passenger railway. The franchise bids of the train operating companies forecast demand growth of nearly 25% in terms of passenger mileage by 2002/03 with the strongest growth in the inter-city market.

3.28 In 1997 we revised the objectives of the Franchising Director to put the passenger first. We welcome the steps now being taken by some operators to put more emphasis on passengers' needs and increased service frequencies, especially where this reduces overcrowding and encourages new passengers. The benefits of our new approach are already beginning to show. For example, the '**passenger dividend**' from Thames Trains includes station improvements, a new Oxford-Bristol service and new bus/rail and bike/rail integration.

3.29 Faster journey times can encourage greater use. That is why we welcome improvements such as the modernisation of the West Coast Main Line. Together with the up-grading there is the commitment to allow for faster, tilting trains, to which the Virgin Rail Group is guaranteeing substantial investment. Such improvement programmes can produce greater reliability and shorter journey times, thereby making rail a more attractive transport option.

> **Fare choice**
>
> - LTS has offered 25% off the weekly peak time ticket price on "Early Bird" trains from selected stations between Southend and London. Fares have been reduced for passengers travelling between 6.30am and 7am – encouraging commuters to switch from their cars when there is the capacity to carry them quickly and comfortably into the City;
>
> - Chiltern Railways has an easy payment plan that spreads the cost of an annual season ticket over ten monthly direct debit payments.

3.30 The ability of the railway to cope with the increase in passenger demand that we wish to see will depend in part on the pace of infrastructure works and rolling stock improvements. Some inter-city routes can increase rail capacity substantially at relatively short notice and at moderate cost, using longer trains and platforms, more trains and improved signalling. Other operators are constrained by infrastructure pinch-points that are already operating at or close to capacity.

3.31 Railtrack has recently identified 15 key bottlenecks on the rail network, together with possible solutions, in its 1998 Network Management Statement (see map at Annex F). Railtrack is evaluating these pinch-points and estimates that its programme for solving these congestion problems could be complete by 2006. **We welcome the Rail Regulator's examination of Railtrack's Statement against the obligations in its licence.** In particular, he is investigating the sufficiency of:

- Railtrack's commitments to improved day-to-day performance of passenger and freight services;

- committed plans to deal with bottlenecks on the network;

- committed projects to renew and develop the network;

- committed plans to meet the requirements of freight.

MORE RAIL FREIGHT

3.32 We can move more freight by rail, relieving pressure on the road network and bringing environmental benefits. The main rail freight operator, English, Welsh and Scottish Railway (EWS), has an aspirational target of doubling its traffic measured by tonne-kilometres over five years and tripling it over ten. Freightliner, which specialises in the haulage of containers between deep sea ports and inland terminals, aims to increase the volume of containers carried by 50% over five years.

3.33 We endorse these targets. Overall, reaching them could mean that in 2010 the share of freight going by road[3] was 10% lower than is currently forecast. **For every percentage point reduction in road freight that is achieved some 1,000 to 2,000 heavy lorries could be taken off our roads.** But we also wish to see improvements to the passenger railway, which must be balanced against the needs of freight customers. We will therefore ask the *Strategic Rail Authority* to develop targets for both the freight and passenger railway in order that we secure the maximum benefit overall from our rail network. In the meantime, we will continue to work towards our objective of moving more freight by rail and towards the targets set by the industry.

> **Rail freight starting to grow**
>
> - 277,000 tonnes of steel products switching from road to rail, with up to an additional five trains per week from Llanwern in south Wales to Wolverhampton Steel Terminal;
>
> - tenfold increase in wagonload business (Enterprise service) between 1994 and 1997. New 75 mph Anglo-Scottish service five days a week;
>
> - new flows of palletised goods for supermarket and chemist chain stores;
>
> - new traffic within the last year from ports such as Workington, Boston, Ipswich, Goole, Immingham and Hull;
>
> - operators and local authorities discussing potential traffic involving ports such as Great Yarmouth, Lowestoft and Kings Lynn.

3.34 We have already made a start on helping to create the right conditions for the revival of rail freight. We have more than doubled expenditure on freight grants. We have negotiated with the French Government and Eurotunnel arrangements to ease access of rail freight through the Channel Tunnel and beyond. Our concordat with the Rail Regulator emphasises the importance of promoting rail freight; the Regulator has secured the creation of extra rail freight capacity on the West Coast Main Line as part of his consideration of plans for a major passenger upgrade. Looking ahead, the *Strategic Rail Authority* will ensure that freight is given proper consideration in the operation and planning of the network; and to the obstacles to growth, as highlighted by EWS in its evidence to

---

3 measured in tonne-kilometres

the Environment, Transport and Regional Affairs Committee[4], and which include loading gauge, track capacity constraints, and access to additional land.

3.35 We will issue revised planning guidance (see Chapter 4) to facilitate more freight to be moved by rail. Local authorities in preparing development plans will be expected to consider, and where appropriate protect, opportunities for rail connections to existing manufacturing, distribution and warehousing sites adjacent or close to the rail network and allocate sites for suitable new developments which can be served by rail.

Rail freight: containers on the move at Felixstowe.

BETTER LOCAL RAILWAYS

3.36 In drawing up *local transport plans*, local authorities will take account of the potential contribution of rail (both conventional and light rail) to their strategies for reducing car use. The potential is likely to vary significantly between different types of authority and whether they serve urban or rural areas.

3.37 Light rail, and similar rapid transit systems, can have a role to play in delivering integrated transport in urban areas – particularly if planned as part of an overall strategy. The capital costs of light rail systems are, however, high – particularly in comparison to bus priority measures and more modest guided bus schemes which may offer a more cost-effective alternative.

Greater Manchester Metrolink

- runs mainly along an old heavy rail corridor replacing two heavy rail services (Altrincham to Manchester Piccadilly and Bury to Manchester Victoria) providing a rail link into and through the city centre;

- at a cost of £150 million (around one third of which would have been required to keep existing rail lines open), it carries 14 million passengers a year;

- passenger numbers are up on the old heavy rail and there is clear evidence of some switch from car use;

- Metrolink, owned by the Greater Manchester Passenger Transport Executive was built under a Design Build Operate Maintain contract. Altram (Manchester) Ltd, a private sector consortium operates the system and will operate the extension to Salford Quays and Eccles due to open by the end of 2000.

3.38 In due course, we shall expect local authorities wishing to develop light rail systems, to use revenues from new congestion charging schemes or parking levies as a source of funding for such systems (see Chapter 4). In the meantime we believe that resources available for funding local authority capital expenditure on transport can, in general, be used more productively supporting packages of more modest measures which spread benefits more widely. Funding for new major light rail schemes will therefore not be a priority and schemes will be supported only if they represent good value for money and form an integral and necessary part of a strategy in a *local transport plan* – demonstrating clearly that the objectives of the plan cannot be met in alternative ways. We would also expect local authorities to develop public-private partnerships to take forward such schemes wherever it is sensible to do so.

---

4   Third Report of the Environment, Transport and Regional Affairs Committee, House of Commons Session 1997-98, on the proposed Strategic Rail Authority and Rail Regulation, HoC paper 286-1, March 1998

CHAPTER 3 Integrated Transport

> **Women and Transport**
>
> Women's transport needs are often different. Although they make about the same number of journeys on average as men, these are shorter and they walk and use public transport, especially buses, more. Men are more likely to have first call on the car in a one car household. Many women have concerns about their personal security, particularly when on their own and at night.
>
> Our *New Deal for transport* will mean for women:
>
> - greater emphasis on integrated transport, including more accessible buses, better information and safer interchanges;
> - safer public transport, including the Secure Stations Scheme;
> - improving the quality of the pedestrian environment, eg making it easier for women with children in prams to get about;
> - land use policies to encourage local services, reducing the need to travel by car;
> - women's transport needs to be assessed in *local transport plans* and through auditing transport initiatives;
> - safer routes to school initiatives;
> - *Commission for Integrated Transport* to take full account of women's transport needs.

Better for the motorist

3.39 The *New Deal for transport* gives people more choice about when and where to use their car. We want to ensure the alternatives are real and attractive. Our new approach is about widening choice, not forcing people out of their cars when using a car is their preferred option. Our transport system does not always serve the motorist well, as has been underlined by motorists' organisations in their responses to our consultation. This is why we are introducing a *New Deal for the motorist*. The *New Deal* includes more reliable journeys from better maintained roads, improved management of the network and a targeted programme of investment.

3.40 We want to see more opportunities for cars to be used as part of an integrated transport system. We are therefore encouraging park and ride facilities to town centres to help beat congestion and at railway stations to make for easier long-distance travel.

Moving freight

3.41 We are extending choice to all users of transport. Our proposals for sustainable distribution, described later in this Chapter, will provide greater choice for moving freight, to promote a more efficient industry and a better environment. There will be new opportunities for distribution in town and city centres, as vehicles become quieter and cleaner. *Quality Partnerships for freight* will produce local solutions to local problems. We will promote the role of rail freight, inland waterways and coastal shipping in the movement of goods, providing a real alternative to moving freight by road.

Getting to the airport

3.42 We set out later in this Chapter how the *New Deal for transport* will make it easier to get to airports by public transport. People should not have to use their car for journeys when they don't want to.

3.43 We want to see more choice for passengers in a way that improves the environment. Our proposals to encourage international access to regional airports will reduce reliance on the London airports, cut the length of surface journeys and increase convenience for the passenger.

The role of motorcycling

3.44 Mopeds and motorcycles can provide an alternative means of transport for many trips. Where public transport is limited and walking unrealistic, for example in rural areas, motorcycling can provide an affordable alternative to the car, bring benefits to the individual and widen their employment opportunities.

3.45 Whether there are benefits for the environment and for congestion from motorcycling depends on the purpose of the journey, the size of motorcycle used and the type of transport that the rider has switched from. Mopeds and small motorcycles may produce benefits if they substitute for car use but not if people switch from walking, cycling or public transport.

3.46 The role of motorcycling in an integrated transport policy raises some important and complex issues. We are therefore setting up an advisory group to bring together motorcycle interests and other interested parties. This will allow discussion of issues of concern to those who ride motorcycles and of the ways we can work together on policies, including encouraging further improvements in the safety and environmental impact of motorcycling.

3.47 In drawing up their *local transport plans*, local authorities should take account of the contribution that motorcycling can make and consider specific measures to assist motorcyclists, such as secure parking at public transport interchange sites. We would welcome proposals from local authorities interested in conducting properly monitored pilot studies of the use of bus lanes by motorcycles, to help inform decisions on whether there is a case for motorcyclists to be allowed to use bus lanes.

More integrated public transport

In pursuit of the seamless journey

3.48 For public transport to provide an attractive alternative to the convenience of a car, it must operate as a network. With the *New Deal for transport* there will be:

- more through-ticketing;
- better facilities at stations and other places for interchange;
- better connections between and co-ordination of services;
- wider availability and provision of information on timetables, route planning and fares;
- a national public transport information system by 2000, available over the telephone, internet etc.

3.49 In preparing their *local transport plans* local authorities will be required to address these matters. For the most part, improvements can be gained through the co-operation of public transport providers and through effective partnerships with local authorities. But, where necessary, we will strengthen local authorities' powers to secure integration.

3.50 The integrated transport system that we want can already be seen in other parts of Europe; for example:

- fares and ticketing – in the Netherlands, 'strippenkaart' tickets allow passengers to make a fixed number of journeys in different Dutch cities using any type of public transport;

- interconnecting services – in Hanover and Stuttgart, evening passengers can ask their tram driver to radio ahead for a taxi to meet them at their destination stop. The cost is included in the tram fare as a flat-rate add-on;

- passenger information – since May 1992 passengers in the Netherlands have been able to ring a single national telephone number for full door-to-door timetable, fares and other information;

- interchange facilities – vandal-proof lockers for cycle storage are provided at stations in the Netherlands. Local buses in Basle and Tubingen carry bikes on special racks or platforms.

Fares and ticketing

3.51 Tickets which are easy to get, offer value for money, flexibility and make changing easy can encourage more people to use public transport

3.52 Rail operators are required to offer through-ticketing for all rail journeys. There have been some problems but the requirement is closely monitored by the Rail Regulator. There are no equivalent obligations on bus operators. We welcome the positive action taken by some companies to accept other operators' tickets or participate in area ticketing schemes, but more needs to be done. We also welcome the increasing number of operators who are starting to introduce initiatives such as rail-bus tickets. We will encourage their wider use. We want to see more 'travelcard' schemes across the country.

3.53 Local authorities when preparing their *local transport plans* should consider the arrangements for through-ticketing and travelcards. We will publish guidance on good practice and ensure that the necessary powers are available locally to require operators to promote and participate in joint-ticketing/travelcard schemes.

> **London Travelcard**
>
> - one of the best examples in Britain of a successful area ticket scheme;
>
> - provides unlimited pre-paid travel within specified zones on bus, rail, underground and Docklands Light Railway services throughout the capital;
>
> - London Transport estimates that introducing the Travelcard increased bus passenger miles by one fifth with underground use going up by one third.

3.54 The structure of bus fares outside London can be very complex. This can add to the time buses spend at stops whilst fares are collected. Unnecessary delay at stops makes buses less attractive and adds to congestion. Passengers who are forced to change buses in the course of their journey usually have to pay twice and pay more than they would if the journey had been made on one bus.

3.55 Our proposals to provide local powers to ensure that bus operators participate in multi-operator ticketing schemes will go part of the way to resolving these problems. But we are also looking to the bus industry to introduce simpler fare structures and through-ticketing, where necessary in co-operation with local authorities.

3.56 Technology can help to provide better ticketing arrangements. The most should be made of smartcards. We are reviewing the capabilities of technology with key players in the industry, both to identify the potential benefits for integrating journeys and to see what role Government should play to help bring forward viable applications.

More integrated public transport

Integration through technology

- CONCERT research project, supported by the European Commission, is piloting integrated payment systems using smartcards for parking fees, bus and rail fares, and in some cases road tolls;
- pilots in Marseilles, Bristol, Bologna, Dublin, Barcelona and Trondheim.

3.57 We will encourage all bus and rail operators to offer carnets (batches of single rail or bus tickets bought at discounted rates) as part of their ticketing range. They can be a flexible alternative to season tickets for part-time workers, and useful for the occasional traveller.

Physical interchange

3.58 Many journeys include an interchange, from the relatively straightforward change of buses at a bus stop to major rail stations and airports where several ways of travelling come together. Quick and easy interchange is essential for public transport to compete with the convenience of car use, which is why we will expect local authorities' *local transport plans* to consider interchange facilities. These audits will assess the adequacy of existing facilities against the key attributes of good interchange:

- reliable/punctual and frequent services to produce minimal waiting times;
- short walking distances and clear directional signs;
- good timetable displays;
- staff availability;
- well maintained infrastructure, including public conveniences and baby changing facilities;
- good personal security;
- accessibility.

Good practice for interchange

- Sheffield, Leeds and Laganside in Belfast are examples of high quality bus stations that have been opened in recent years. They have smart and clean waiting facilities, with electronic passenger information systems, travel enquiry centres, retail outlets and security arrangements;
- cycle lockers to provide secure and weatherproof storage are being introduced at Ipswich railway station and stations in the West Yorkshire PTA area. Some operators, including Anglia Railways are installing cycle racks on trains;
- Oxford park and ride is the largest in the UK and an essential element in holding down traffic levels in the city.

3.59 Local authorities will be expected to identify the improvements that need to be made. Funding will be available through *local transport plans* for improving interchanges – especially to help disabled people and for pedestrian and cycle access. We will encourage greater use of public-private partnerships to fund improvements.

3.60 Designing for better interchange can yield significant benefits and represents good value for money. For example, many towns have re-organised their high street bus stops and now have groups of stops served by interconnecting services. Small scale improvements which can make a real difference but which are often overlooked include:

- better protection from the weather;
- instantly readable and relevant information on routes and frequencies;
- better directional signs between, for example, bus stops and between rail and bus stations;
- regular cleaning and maintenance;
- secure parking for bikes at bus shelters.

3.61 We will commission further research[4] in order to update guidance on interchange, identifying best practice and good design. The guidance will cover the needs of disabled people and will consider the planning process. It will look at the way shops and cafes, well-maintained toilets and baby-changing facilities, and attractive architectural design and public art can add to quality of interchanges and make them safer and more inviting places.

Better interchanges: the award winning Birkenhead bus station.

Integration in action on Anglia Railways.

3.62 Pedestrian access to rail and bus stations is often poorly designed and can be hazardous. Significant measures also need to be taken to improve provision for cyclists. This is relatively limited even at the larger rail stations and where storage facilities are provided, security is often poor, deterring cyclists from using trains and rail passengers from cycling to the station. We will look carefully in our additional research into interchange at how pedestrian and cycle access can be improved.

3.63 All rail operators will be asked to report on their success in meeting the objectives in the code of practice for rail operators developed by Sustrans, the Cyclists Public Affairs Group and the CTC. We will collaborate with local authorities, public transport operators and other bodies to help establish acceptable methods of carrying cycles on buses and coaches.

Providing for cyclists, Sustrans' code of practice for rail operators

Rail operators should provide as far as is reasonably practicable:

- general customer information on cycle facilities;
- improved access for cyclists to stations;
- sufficient, adequate and convenient cycle parking at stations – under surveillance and well-signed;
- onboard storage of bicycles which is sufficient, safe and secure and does not unduly inconvenience other users;
- at-station information and help for cyclists.

3.64 Local development plans should consider allocating sites for interchange; for example, for park and ride to town centres and at bus and rail stations. Local planning authorities can protect these proposals through the exercise of their development

---

4   building from "Transport Interchange – Best Practice", Colin Buchanan and Partners, 1998.

control responsibilities. To help local authorities we have commissioned research into what makes park and ride successful and its effect on car mileage. On completion of the project, which is expected shortly, we will publish advice on best practice.

Timetable co-ordination and service stability

3.65 We will bring forward changes to promote service stability and limit the frequency of bus timetable changes as well as improving the quality of timetable information. These will include changing the period of notice required before a registration with the Traffic Commissioners, or its variation, becomes effective, introducing set dates for service changes and proposals for requiring operators to provide service and schedule information electronically in a standard format. Some of these changes can be made by secondary legislation after consultation but others will require primary legislation.

3.66 Our proposals will, of course, reflect the need for operators to retain sufficient flexibility to make essential and timely adjustments to meet passenger demand. Bus operators and local authorities will be expected to make progress on a voluntary basis in the interim.

3.67 The current railway performance regime – the incentives system used by Railtrack and operators – could be improved to encourage train operators to hold connecting trains when delays occur. We look to the Rail Regulator to address the weaknesses of the current system.

3.68 We will continue to encourage bus and train operators to develop the potential of integrated bus and rail services. Some train operators already operate feeder bus services linking stations to those towns that have no rail routes or inadequate connections. We expect the pace of these initiatives to accelerate with increased co-operation between bus and train operators. We will issue general guidance on the application of the prohibitions in the Competition Bill, so as not to deter co-operation between operators that is in the interests of connecting services, co-ordinated timetables and integrated networks.

3.69 Local authorities will be expected to establish groups with transport operators, user groups and others to discuss timetable needs and planning. Their recommendations will inform the preparation of *local transport plans*.

Passenger information

3.70 Although operators have recently improved passenger information, its quality still varies dramatically across the country. It is quite good for rail journeys, variable for bus journeys and only good in a few places for journeys involving bus and rail.

For journey planning the customer needs information on

- timetabling;
- services;
- fares;
- interchange details and facilities;
- how to book;
- delays and engineering works.

3.71 Train operators are required to co-operate in the provision of passenger information and information must be impartial between rail companies. As part of their licensing agreement, train operating companies are obliged to provide timetable and fare information for a central database[5] and operate the National Rail Enquiry Service collectively. The Rail Regulator is presently responsible for enforcing licence conditions

---

5   Railplanner contains information from Railtrack's database

and has been active in doing so. Train operators and Railtrack are now working together to improve information. The improvements include:

- common standards for information displays and timetable information;
- development of 'real time' information for passengers;
- co-operation between operators following service disruption.

3.72 There is no obligation on bus operators or local authorities to provide published timetables but most of them do so. Local authorities often provide area-wide timetables derived from the information that operators are obliged to send them when registering services with the Traffic Commissioners. There are good examples of well-designed information backed up by telephone enquiry points.

> **Great Britain Bus Timetable**
>
> - published by Southern Vectis three times a year. Provides comprehensive coverage of long distance bus and coach services, with limited coverage of local services;
> - Southern Vectis also operates a central telephone enquiry service, providing telephone numbers for individual operators so that more detailed information about services (including rail services) can be obtained.

3.73 Getting timetable and connection information is vital for many passengers. We are keen to see a national integrated journey timetable set up. The best way forward is to develop a framework which builds on information already available[6] and draws on new information schemes as they become available. Passengers would access the system through one enquiry point, even though information would be drawn from different sources. The enquiry points could include a telephone information line, enquiry bureau, teletext and the internet.

3.74 In partnership with local authorities, operators and user groups, we will seek agreement on the format of information and interfaces between different systems, and co-ordinate research to provide both local and national coverage. **Our aim is for a public transport information system to be systematically extended across the country by 2000.** The initial focus will be on timetable information but the framework will be developed with the aim of including information on fares.

3.75 We will also develop our existing guidance on passenger information including timetables, fares, interchange and booking information across all types of public transport and different media. The new guidance will in addition cover the marketing, promotion and presentation of information, and best practice for in-journey information.

3.76 We will ensure that local authorities and transport operators are aware of their duties under Part III of the Disability Discrimination Act which will require them, in certain circumstances, to produce information in formats which are accessible to disabled people. This might include information provided in large print or on audio tape for visually impaired people or given via a minicom for people who are hard of hearing. We will announce the timetable for implementation of the remaining duties in Part III in due course.

---

6 "Review of Telematics Relevant to Public Transport", Transport Research Laboratory, 1998.

More integrated public transport

Using new technology

- London Transport's ROUTES (Rail Omnibus Underground Travel Enquiry System) is a sophisticated information system, providing real time multi-modal information on travel in the Greater London area;

- North West Trains has a web site providing real time information on the state of arrivals and departures at stations in its area as well as timetable and journey planning information;

- Buckinghamshire County Council provides a comprehensive county-wide public transport guide with a linked map, timetable and route finder;

- Southampton and Winchester have bus arrival time information at bus stops and late running buses are given priority at junctions through the SCOOT traffic signal control system;

- Tyne and Wear is developing a transport information service using teletext on cable television and the internet aimed specifically at elderly and disabled people.

3.77 To help secure improvements in passenger information at the local level, we will require local authorities to ensure that information about bus services is available in their areas, including at bus stops. This will enhance local authority involvement in promoting public transport. Local authorities will have new powers to secure the availability of passenger information where necessary and to recover the costs from operators. These changes will require primary legislation.

3.78 In the short term, we intend to introduce a series of small scale improvements via secondary legislation. These will strengthen the requirements on bus operators to display timetables and fares inside buses.

Better taxis

3.79 Taxis are an important part of an integrated public transport system and, together with private hire vehicles (PHVs), fill the gap when most buses and trains have stopped for the night. Local authorities will need to consider these vehicles in their *local transport plans* including, for example, the priority they are to be given when road space is reallocated and whether there are sufficient taxi ranks in the right places, operating at the right times of day.

Taxis can be an important link.

3.80 It is important that local authorities use their taxi and licensing powers to ensure that taxis and PHVs in their district are safe, comfortable, properly insured and available where and when required. Outside London, taxis and PHVs are regulated by local authorities to check that vehicles are safe and that drivers do not have relevant criminal convictions. In London, the taxi trade is regulated but there is no criminal record check of minicab drivers, nor proper checks on the vehicles or minicab companies. Inadequately regulated minicabs are open to abuse and at worst are an unsafe way to travel. Therefore, following consultation last year, we have concluded that there should be regulation of London minicab drivers, vehicles and operators. We are supporting a Private Member's Bill on this matter but if that should fail for any reason then we would introduce a Bill as soon as Parliamentary time permits.

Travelling without fear

3.81 Many of the responses to our consultation for this White Paper suggest that concern about personal security is a constraint on the use of public transport and walking. This can be worse at night and for older people, women and ethnic minorities. People who live in inner city areas with high crime levels can suffer most. Research[7] has suggested that over 10% extra patronage of public transport could be generated mainly in off-peak times if travellers, particularly women, felt safer in making their journeys. There is a virtuous circle here – fuller trains and buses make people feel safer when travelling.

> **Government's objectives for the police**
>
> - targeting and reducing local problems of crime and disorder;
> - making towns and neighbourhoods safer will help promote walking, cycling and public transport as alternatives to the car;
> - securing co-operation of all, including the local community.

3.82 The *New Deal for transport* is about giving people choice. We want people to make more use of public transport but understand that for some, especially women, and for some time, the private car will continue to be perceived as providing the safest way (in terms of personal security) of getting around. **The reduction of crime, and fear of crime, wherever it occurs in the transport system will be a major priority.**

3.83 We will work with local authorities, transport operators, the police and motoring and other organisations on specific measures to reduce fears about personal security on transport, and more generally in the planning and design of urban and rural areas.

> **Tackling car crime**
>
> We are working across Government with the car industry and insurers, motoring and consumer organisations and the police to reduce vehicle crime by, for example:
>
> - revamping and relaunching the secured car parks scheme, promoted by the Association of Chief Police Officers with the support of the Home Office and administered by the AA. This aims to create a safer parking environment by increasing the number of accredited car parks from 450 to 2,000 by the year 2000;
> - analysing and publishing vehicle crime data to inform motorists of the risk of theft by make and model;
> - setting targets for manufacturers on the performance of vehicle perimeter security and immobilisation devices.

3.84 We will encourage the spread of best practice in crime prevention techniques on public transport. In particular, we will identify and evaluate current crime prevention initiatives and issue guidance on good practice measures to improve security for passengers and pedestrians.

3.85 There are already initiatives where train operators are working with local authorities to improve security at stations. We welcome these, not least because stations are a key area of concern for lone travellers, particularly women. Some operators are reinforcing security on their trains through, for example, the use of CCTV.

3.86 We expect all public transport operators to adopt the crime prevention strategies contained in our guidance "Personal Security on Public Transport – Guidelines for Operators". Simple measures can be important, for example, better lighting and training and availability of staff.

---

7 "Perceptions of safety from crime on public transport", Crime Concern and Transport and Travel Research, 1997.

Station staff also have an important role in helping their customers, particularly elderly and disabled people, to use services.

3.87 We have recently launched, with the British Transport Police and Crime Concern, the "Secure Stations Scheme" aimed at fighting the fear of crime at stations. Further measures may also be needed to make car parks near stations or at park and ride sites even safer, to encourage more people to use public transport for part of their journey.

> **Secure Stations fight fear of crime**
>
> Under the new "Secure Stations Scheme" all 3,000 stations policed by the British Transport Police can apply to become Secure Stations. The scheme establishes the first ever national standards for station security. To be accredited, stations must meet management and design standards for:
>
> - trained staff and close-circuit surveillance;
> - rapid response in emergencies;
> - regular inspection and maintenance;
> - better lighting and secure fencing;
>
> Standards apply to station platforms, interiors, approaches and car parks.
>
> Station operators have to conduct an independent passenger survey to see whether passengers feel safe at the station and provide evidence of low crime rates over a sustained period.

3.88 Transport staff also deserve to be free from the fear of crime. We will encourage good practice by all public transport operators to protect their staff. The Department of the Environment, Transport and the Regions' (DETR) practical guide on "Protecting Bus Crews" sets out measures to reduce the risks for bus crews.

3.89 For bus passengers, the greatest fear about personal security is waiting at the bus stop and on the walk to and from the bus stop at either end of the journey. This is something that can be tackled in part through getting street design right in the first place, as well as by enhanced security through measures such as CCTV – which also has a part to play in making bus journeys feel safer.

3.90 Attention also needs to be given to the design and layout of bus stations and their operation, particularly at night, in order to increase passenger security. Revised planning guidance in England (see Chapter 4) will highlight the need for environments that are convenient, attractive and safe for walking.

3.91 Concern for personal security can also impose extra costs. For example, people preferring to travel in pairs or in groups may have had to look to taxis as a cheaper option than public transport. But there are alternatives. We will encourage marketing schemes such as 'two for the price of one' which can help to keep people using public transport, particularly after dark.

> **Encouraging group travel**
>
> - South West Trains 'family fare' – 1997 Christmas promotion allowed five people to travel to Guildford from local stations for a total of £5;
>
> - Centro (West Midlands PTE) daytripper – up to six people benefit from the daytripper card for no more than the price of one adult and one child;
>
> - 'Kids for a Penny' – in a bid to encourage family bus travel, Trent Buses ran a scheme last summer (June to August) allowing a child accompanied by a paying adult to travel on their buses for just 1p.

Accessible transport for disabled people and easier access for all

3.92 Public transport must meet the needs of all in our community and 'accessible' public transport is vital for disabled people in particular, so that they have the opportunity to play a full part in society. The steps we are taking through the Disability Discrimination Act will mean that in future public transport is accessible to disabled people as a matter of course, including those who need to use a wheelchair. This will also make life easier for the growing population of people who are elderly and those who need to travel with a baby-buggy or pram, or heavy shopping.

3.93 We are bringing into effect the requirements for new rolling stock on the railway from the end of this year. For buses and taxis the implementation dates are being set following consultation. We have consulted on an implementation date of 1 January 2002 for taxis and a range of dates according to different bus and coach types, starting with 1 January 2000 for large single deck vehicles.

3.94 From 1 January 1999, to conform with EU law, we will raise the maximum axle weight of buses and coaches from 10.5 to 11.5 tonnes and increase their maximum gross weight from 17 to 18 tonnes. We will bring forward the necessary legal changes shortly. This change will allow some safety and accessibility improvements to buses and coaches such as the ability to design low-floor buses, without imposing significant reductions in carrying capacity.

3.95 Accessibility is a much more complex issue than simply making it easier to get on and off public transport. To get the most out of investment in accessible public transport, local authorities and transport operators will have to consider the needs of disabled people from start to finish of their journey. This involves tackling barriers in the street, at bus stops and at public transport interchanges. The availability of staff to help disabled people is important.

3.96 Local authorities can use the land use planning system to ensure that developments are accessible to disabled people. When drawing up their *local transport plans*, local authorities will be expected to address accessibility issues. We will draw up guidance to help them.

> **Widening choice**
>
> Europa Buscentre and Great Victoria Street Railway Station, Belfast:
>
> - an integrated bus and rail facility in a fully accessible environment;
>
> - design features include low level counters at booking offices, low level public telephones, textphone facilities, tactile flooring, high contrast signage, an induction loop, parents' room and toilets for disabled customers.
>
> Buchanan Bus Station, Glasgow:
>
> - provides level access throughout with automatic doors and dropped kerbs;
>
> - has low level telephones and wheelchair accessible toilets and a wheelchair is available on request for people with walking difficulties.

3.97 Because the accessibility regulations under the Disability Discrimination Act will apply only to new buses, coaches, trains and newly licensed taxis, it will take time to achieve a fully accessible transport network. Good progress is already being made by the bus industry in introducing modern, accessible buses into the fleet. Some local authorities have introduced grants to prompt operators to try low-floor buses by 'topping up' the difference in cost compared with a

More integrated public transport

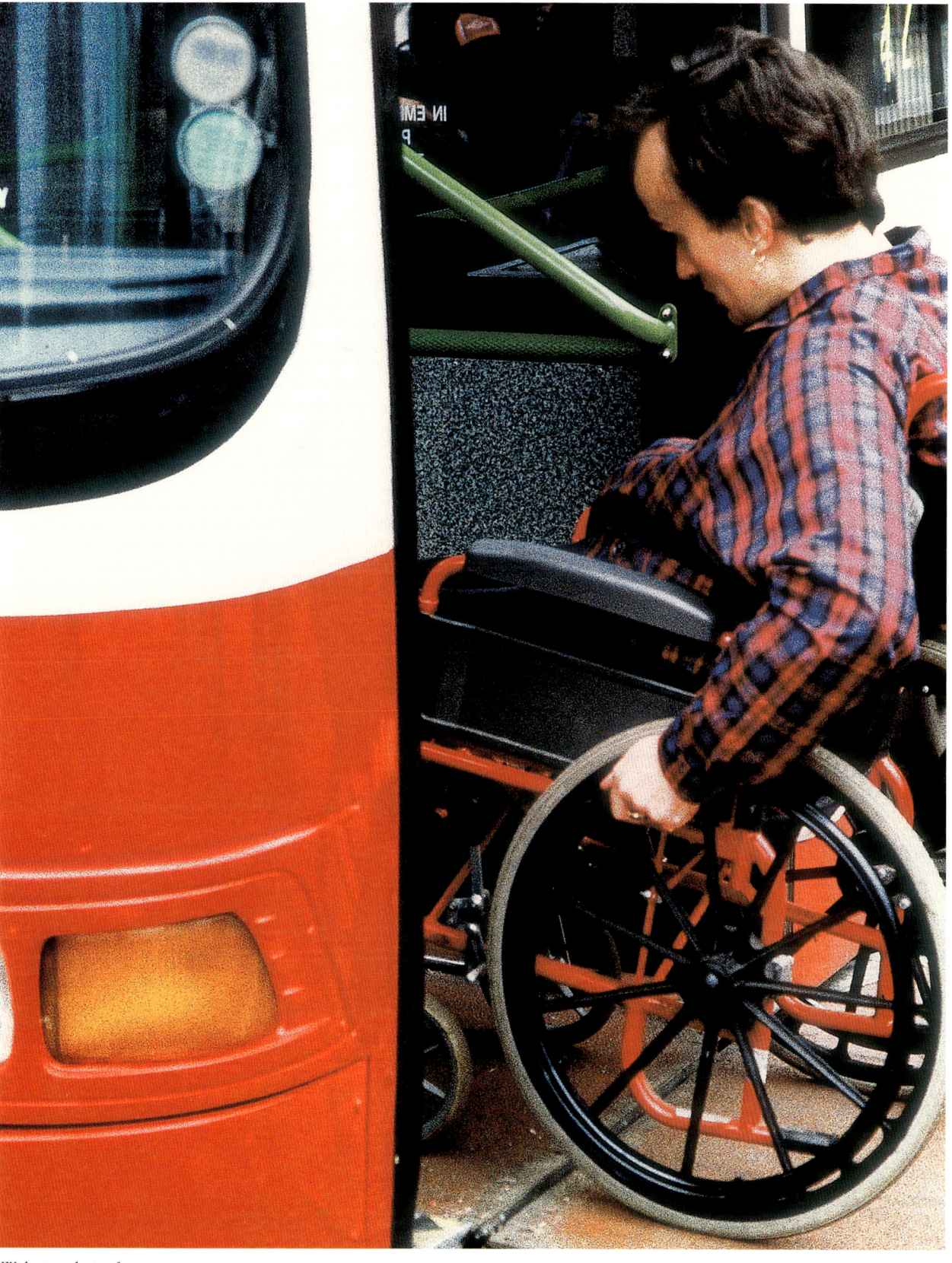

Widening choice for everyone.

conventional bus. We expect our proposals for Quality Partnerships to accelerate the introduction of low-floor buses.

3.98 On the railway, much can be achieved within the existing regulatory framework. For example, all inter-city services are fully accessible and new services such as the Heathrow Express are designed and built to offer full access. The Rail Regulator has a duty to take account of the needs of disabled people. This includes the production of a Code of Practice, which has been drawn up in consultation with the Disabled Persons Transport Advisory Committee. In addition, the *Strategic Rail Authority* will have a specific duty to protect the interests of disabled people and promote the provision of accessible transport.

3.99 We want airports and ferries to be more accessible and cater for the needs of disabled people. We also want more taxis to be accessible to disabled people and for private hire companies to make greater efforts to respond to their needs. But we appreciate that for some, severely disabled people in particular, a car may be the only viable way of getting around. The *New Deal for transport* is about widening choice not forcing people out of their cars. **Anyone who meets the required standards will have the right to hold a driving licence and own and use a car.** We have already, for example, exempted vehicles first registered in the 'disabled exempt' tax class from the fee that was introduced in April for the first registration of a vehicle. Disabled people registering their vehicle in this class are among those with the most severe mobility difficulties.

Streets for people

Integration on local roads

3.100 Through the *New Deal for transport* we will improve the environment in towns and cities and create the conditions for people to move around more easily. More road space and priority will be given to pedestrians, cyclists and public transport.

3.101 We will achieve this by a different approach to traffic management. This new approach will also help to achieve the air quality objectives of the National Air Quality Strategy. Local authorities will be expected to take a strategic view of traffic management when preparing Regional Planning Guidance and development plans (see Chapter 4), considering how different measures can complement each other. *Local transport plans* will set out how these measures are to be delivered at the local level.

Making better use of local roads: vehicles carrying more than one person benefit from this priority lane in Leeds.

Local traffic management: the potential

Bus priority

- significant scope for development in larger towns, with traffic restraint measures;
- local transport plans to develop and implement coherent and comprehensive policies;
- Quality Partnerships and *Quality Contracts* to secure better bus services.

Traffic calming

- scope for development of new designs of traffic calming in, for example, historic cores of some towns, popular countryside destinations and rural lanes; low speed and home zones in residential areas.

Priority routes

- cycle route networks;
- pedestrian route enhancements;
- priority route networks as in London and Edinburgh provide a framework for application of traffic management policies, eg bus priority, parking restraint, urban traffic control.

Urban traffic control

- early progress possible in local authorities to make fuller use of the best facilities already available;
- over time, Government/industry collaboration on new range of modern urban traffic management systems.

Driver information

- good signing can help efficient use of the network. It needs to be well-maintained and updated; signing can be made less environmentally intrusive;
- new techniques such as automatic incident detection offer the prospect of strategic traffic management control of highway networks;
- use of in-vehicle information services likely to grow; route guidance will help to reduce unnecessary travel, especially when live traffic information is incorporated.

Vehicle measures

- restriction of certain areas to 'clean' or 'quiet' vehicles.

Parking

- control of on-street parking to prevent vehicles obstructing traffic and pedestrians;
- new types of equipment for controlling on-street parking; electronic meters, pay and display machines operated by magnetic cards, and voucher systems;
- parking enforcement by local authorities, penalties used to fund enforcement, scope for more authorities to take up new powers;
- parking control, on and off-street, as a component of plans to reduce the amount of travel in and to congested town centres;
- parking restraint strategies that include packages of measures to improve access to town centres by public transport and deter through-traffic and a levy on parking at the workplace can substantially reduce the amount of traffic in central areas;

Car Sharing Lanes

- High Occupancy Vehicle Lane in Leeds recently opened as part of EU research project, will be monitored for progress and potential elsewhere.

3.102 Local authorities should not have to 'reinvent the wheel' in traffic management. We will provide advice and guidance, and disseminate the principles of good practice that emerge from our traffic management research programme. We will also encourage the use of new technology in traffic management where appropriate and cost-effective.

3.103 In the past there has been some concern that a different approach to traffic management could cause excessive congestion on other parts of the network. Research[8] suggests that this concern can be exaggerated and has stressed that schemes should be judged against a broad range of objectives. We will encourage the development of new appraisal systems that take account of the wider benefits of a more radical and comprehensive approach to traffic management.

3.104 We wish to reduce the impact on traffic and pedestrians caused by street works for utility companies. We will consult on options for an incentive system, with penalties, to minimise disruption to all road users, and to encourage improved co-ordination of streetworks.

People before traffic: shoppers in Cambridge.

Living town centres

3.105 Thriving town centres are the focus of urban life. They are central to sustainable development because they are easily accessible by a choice of transport. Good public transport is essential and so, too, is the quality of environment. People want well-planned town centres where they can live, enjoy shopping, working and local culture. Too often, town centres have been sacrificed to busy roads: the *New Deal for transport* will give priority to people over traffic.

Putting people first in Edinburgh

- road space priorities have been changed with clear benefits for pedestrians, bus users, local business and the environment;

- in the historic Royal Mile, space for pedestrians has been increased substantially and it is closed to vehicles at the busiest times during the International Festival. It is estimated that improvements will lead to an extra £26 million being spent in Edinburgh every year;

- in Princes Street, the main shopping street, traffic levels have been reduced substantially. Accidents are down by a third (14% down in the wider area), air quality is significantly improved and shoppers are spending more.

3.106 Despite initial misgivings from some local traders, pedestrianisation schemes have proved very popular. We will also encourage local authorities to consider traffic calming and the reallocation of road space to promote walking and cycling and to give priority to public transport.

---

8 "Traffic Impact of Highway Capacity Reduction", MVA and ESRC Transport Studies Unit, UCL, 1998.

3.107 We will support local authorities and the haulage industry in the development of 'City Logistics' systems[9], drawing on the experience of projects which have been initiated in other EU countries. These could help to improve efficiency in goods deliveries and reduce pollution.

3.108 We have launched the ALTER project – Alternative Traffic in Towns – during the UK Presidency of the EU to help to produce healthier town centres and cities. ALTER will produce concerted action by cities across Europe to give preferential access in certain areas to vehicles with zero or low emissions. Oxford is one of the lead authorities, together with Athens, Barcelona, Florence, Lisbon and Stockholm. All European cities with a population of more than 100,000 are to be invited to a conference in Florence in October 1998 to take the project forward.

3.109 The concept of 'Clear Zones' is being developed through our Foresight programme.[10] Clear Zones can improve the quality of life in town centres through:

- reducing the impact of traffic while maintaining accessibility, viability and vitality;
- reducing emissions caused by public transport and goods distribution;
- looking at demand management and the provision of efficient interfaces and information between different types of transport.

3.110 A co-ordinator has been appointed under the Foresight Transport Panel to help in the development and demonstration of technologies to achieve these aims, and to support local authorities who wish to implement Clear Zones.

Quality residential environments

3.111 We want towns and cities to be places where people want to live. The *New Deal for transport* will support the urban renaissance that is essential to revitalise urban living and save our countryside from urban sprawl.

3.112 In part, this means people being able to go about their daily business without being intimidated by traffic. Better planning can contribute to achieving better and safer residential environments by influencing the design and layout of new developments. Traffic can be calmed from the outset by designing for low speeds. Sometimes new developments can be designed to be 'car free'.

3.113 In established residential areas we want to see the creative use of traffic management tools. We want local authorities to make greater use of the wide range of techniques now available that allow traffic calming to be introduced cost-effectively and with sensitivity to the environment. This will include more extensive use of '20 mph zones'. In these zones, the frequency of accidents has been reduced by about 60% and accidents involving children have fallen by 67%.

3.114 20 mph zones are most effective in a series of residential streets or other areas, where speeding traffic puts pedestrians, often children, and other vulnerable road users such as cyclists at risk. To encourage greater use, we will issue new guidance. We have already announced proposals to free local authorities to make their own decisions about 20 mph speed limits.

---

9 see section on sustainable distribution below.

10 The Government's Foresight programme aims to encourage business and university scientists and engineers to work together to exploit science, engineering and technology to increase wealth and quality of life. The second round of Foresight will be launched in April 1999.

CHAPTER 3 Integrated Transport

3.115 'Home zones' have been developed in a number of European countries and involve even lower traffic speeds, more pedestrianised areas and design features that emphasise the change in priority to pedestrians and cyclists. They could prove to be a valuable tool in improving the places where people live and children play.

3.116 With good design many of the objectives of homes zones could be achieved within existing legislation. We will welcome proposals by, and work with, local authorities who wish to pilot the idea.

> **Millennium Village, Greenwich – a sustainable urban community**
>
> - school, shops, small businesses, medical facilities, places of worship, community facilities, parkland and open spaces to be within easy walking distance of homes and each other;
> - much reduced car dependency;
> - well connected by public transport through the Jubilee Line Extension to the heart of London and by the Millennium Transit – a modern, low emission and frequent bus service – to local stations.

A more peaceful countryside

3.117 Traffic management can help to produce better and safer local road conditions, both for those who live and work in rural areas and for visitors, and protect the character of the countryside.

3.118 We welcome the Countryside Commission's demonstration projects[11] for traffic management and the support of the local highway authorities concerned. One of the main conclusions of the work is the need for a strategic approach to managing local traffic, otherwise problems are shunted around the countryside from one place to another. Countryside traffic strategies, that enable individual traffic schemes to be brought forward as part of a wider consideration of traffic and transport, will be important parts of *local transport plans*.

3.119 Traditional traffic management measures can have an urban look and can be even more damaging in the countryside than on the appearance of our towns. We will therefore encourage the continued development of new and imaginative ways of designing local traffic schemes to make them more sensitive to their surroundings. The Countryside Commission's work on Village Design Statements and Countryside Design Summaries is a helpful contribution. The Commission is also producing new guidance for traffic management and calming design and last year, with our support, set up the Countryside Traffic Measures Group (CTMG) to spur innovation in rural traffic management. This will broaden our understanding of the way traffic management measures can be designed with sensitivity to the countryside.

3.120 The Countryside Commission will set up later this year a Rural Traffic Advisory Service. It will organise local groups and seminars to speed the adoption of the measures explored in the CTMG and the Commission's research on the design and implementation of rural traffic measures.

---

11 these followed the Countryside Commission's 1992 report "Road Traffic and the Countryside".

"Rural traffic: getting it right", the Countryside Commission's demonstration projects

In Cumbria:

- traffic calming for Crook: responding to a community request;
- Elterwater Parking: relieving parking congestion;
- under Loughrigg: protecting a quiet country lane;
- public transport information;
- cycling services.

In Surrey, "be a star, don't use the car":

- protecting minor roads from 'rat-running' in the Dorking area;
- amenity and safety on the road for South West Waverley;
- congestion in a tourist village: Shere;
- traffic and schools: Lingfield Primary;
- two wheels not four, the Surrey cycleway.

On Dartmoor, "take moor care":

- area speed limit: reducing animal accidents;
- strategic coach route network;
- Okehampton Railway station and interchange;
- village traffic calming schemes;
- cycling schemes: Dartmoor by bike.

3.121 The Countryside Commission envisages working closely with local authorities as part of a 'Quiet Roads' initiative – to introduce measures to make selected country lanes more attractive for walking, cycling and horse riding, in the interests of a more tranquil and attractive rural environment. The Commission is also developing 'greenways' as traffic free routes within the countryside and from towns and cities to the countryside. Together with Quiet Roads they can form networks that provide safe alternatives to car travel.

3.122 We will help the Countryside Commission and local authorities develop these ideas. This could be through advice and support, including regulatory cover for experiment and innovation where appropriate, or by pilot projects linked to rural traffic management. Local authorities will be able to finance such initiatives through funding for their *local transport plans*.

Making better use of trunk roads

Integration

3.123 This White Paper sets a new course for roads policy. The days of 'predict and provide' are over – **we will give top priority to improving the maintenance and management of existing roads before building new ones**. Our *New Deal for transport* means a better managed road network so that it delivers a high quality service to the road user.

3.124 Roads are currently a major source of frustration for drivers, both private and commercial. Parts of the trunk road network are under considerable stress. To tackle this sustainably we need to get all modes of transport and land use planning working together. This is why we made integration one of the five criteria in our review of trunk road policy and of the roads programme we inherited. It is also why it is important that we should bring trunk roads within the regional planning process in England (see Chapter 4). All decisions on road investment will be taken in the context of our integrated transport policy.

Investment strategy

3.125 In the past, the focus of investment has been on building new roads at the expense of managing existing ones. We will change the priority and provide a coherent programme for improving the service offered by trunk roads.[12] We will look at trunk roads in their wider context, and at the part they play in those transport corridors which include road and rail routes. Our priorities for trunk roads will complement improvements to inter-urban travel, by rail in particular, so that they form part of an integrated approach. We will:

- improve road maintenance, making it our first priority. Skimping on maintenance wastes money. If maintenance is delayed too long structural damage is done and much more expensive and highly disruptive repairs are required;

- make the best use of the roads we have already by investing in network control and traffic management measures and in minor improvements. This will include giving priority in specific locations to certain types of traffic, including buses and coaches and heavy goods vehicles;

- promote carefully targeted capacity improvements to address existing congestion on the network, where they support our integrated transport policy.

3.126 Since new roads can lead to more traffic, adding to the problem not reducing it, all plausible options need to be considered before a new road is built. Carefully targeted improvements to existing roads will be considered, generally as part of wider packages including traffic management measures. Traffic calming and measures to reduce traffic will also be considered in conjunction with, and as alternatives to, the construction of bypasses for towns and villages.

Investment criteria

Decisions on when and where to invest in network improvements, including measures to manage traffic, will be taken in the light of the new approach to appraisal based on the criteria:

- integration – ensuring that all decisions are taken in the context of our integrated transport policy;

- safety – to improve safety for all road users;

- economy – supporting sustainable economic activity in appropriate locations and getting good value for money;

- environmental impact – protecting the built and natural environment;

- accessibility – improving access to everyday facilities for those without a car and reducing community severance.

An integrated network

3.127 Trunk roads are an integral part of our transport system. They cannot and should not be managed and developed in isolation. We will manage the trunk road network (and encourage local authorities to manage local roads) as part of a series of transport networks that have good connections between them.

3.128 There are three key aspects to this:

- integration between all types of transport. We want to make it as easy as possible for car drivers to switch to rail, bus and coach by providing good connections between them, by managing roads as part of the wider transport system and by improved co-ordination with public transport operators. This will increase choice and help to create reliable and seamless journeys;

---

12 in England to be developed in the Roads Review Report. There will be separate reports for Scotland and Wales.

Making better use of trunk roads

- integration between road freight and other freight modes. Better connections to rail freight terminals and ports can help encourage hauliers to switch to rail and shipping;

- integration between trunk roads and local roads. The management of trunk and local road networks is already substantially integrated but this needs to be developed further. Both networks cover a range of road types and situations and to some extent the measures for achieving integration on local roads considered earlier in this Chapter will be of relevance to trunk roads.

Making connections: using the Park & Ride scheme, Oxford.

3.129 The likely impact on local roads will be an important consideration in bringing forward traffic management measures on the trunk road network. Through-traffic will be encouraged to use trunk roads, not unsuitable local roads.

A core road network

3.130 The trunk road network varies greatly from place to place, although most trunk roads are of clear national significance. We have identified a core network in England of nationally important routes (see map at Annex E). In defining this network we have taken the following factors into account:

- linking main centres of population and economic activity;

- accessing major ports, airports and rail intermodal terminals;

- joining peripheral regions to the centre;

- providing key cross-border links to Scotland and Wales;

- classification as part of the UK Trans-European Road Network.

3.131 There are a number of trunk roads which mainly serve local and regional traffic. Such roads would be more appropriately managed by the local highway authority, to enable decisions to be taken locally and to be better integrated with local transport and land use planning issues. Our consultation on the strategy for trunk roads in England showed significant support for the 'de-trunking' of these roads. We will consult the Local Government Association and individual local highway authorities in taking forward these proposals for devolving powers.

Making better use

3.132 In England, the Highways Agency is developing a 'Toolkit' of techniques and equipment which can be used individually or in combination for making better use of the network. As well as bringing forward local environmental and safety improvements, we have asked the Agency to focus the development of its Toolkit on:

- integrating the trunk road network with other modes of transport by providing

  – safer and more accessible interchanges between modes;

- clear, comprehensive and up-to-date information using the latest technology to assist route and mode choice;
- priority measures to assist public transport and vulnerable users;

• managing traffic demand on the network, including giving priority to buses, coaches and lorries where appropriate;

• increasing the efficiency of network operation.

> **Giving greater priority to coaches**
>
> • modern coaches can provide a flexible way of filling gaps in the services provided by trains, as well as competing with them on their own merits and in many cases, offering a lower cost alternative;
>
> • M4 Heathrow bus and coach lane – is the first motorway bus lane to come into service. Road space was reallocated to create the dedicated bus lane. Bus journey time and reliability has been improved;
>
> • M4 Junction 3 to Junction 2 – working on proposals for an eastbound bus and coach priority lane.

3.133 Toolkit measures will form part of our approach to making better use of the M25. We will pilot an innovative and imaginative mix of techniques on the M25 that can have wider application elsewhere. The controlled motorway experiment on the western sector of the M25 has already demonstrated that drivers can expect better journeys through smoother traffic flows and a reduction in stop-start driving conditions.

3.134 Toolkit measures are likely to be most effective if deployed as part of a 'Route Management Strategy'. This is a technique being developed by the Highways Agency to provide a framework for managing individual trunk routes as part of wider transport networks. Route management strategies will interlock with local transport strategies (set out in *local transport plans*), within the context established by Regional Planning Guidance.

3.135 In Scotland, the concepts of 'Route Action Plans' and 'Route Accident Reduction Plans' have been in place for several years resulting in the comprehensive study of routes and the application of similar tools to those in the Highways Agency's Toolkit.

The Highways Agency as network operator

3.136 We have set new objectives for the Highways Agency. Overall, the Agency's strategic aim will be to contribute to sustainable development by maintaining, improving and operating the trunk road network in support of our integrated transport and land use planning policies. The Agency's main purpose in future will be as a network operator rather than as a road builder. It will have the following key objectives:

• to give priority to the maintenance of trunk roads and bridges with the broad objective of minimising whole life costs;

• to develop its role as network operator by implementing traffic management, network control and other measures aimed at making best use of the existing infrastructure and facilitating integration with other transport modes;

• to take action to reduce congestion and increase the reliability of journey times;

• to carry out the Government's targeted programme of investment in trunk road improvements;

• to minimise the impact of the trunk road network on both the natural and built environment;

- to improve safety for all road users and contribute to the Government's new safety strategy and targets for 2010;

- to work in partnership with road users, transport providers and operators, local authorities and others affected by its operations, monitoring to promote choice and information to travellers and publishing information about the performance and reliability of the network;

- to be a good employer, managing the Agency's business efficiently and effectively, seeking continuous improvement.

> **The Highways Agency as network operator**
>
> - align trunk road network operation with integrated transport policy;
> - focus on moving people and goods safely and effectively rather than building new roads;
> - optimise use of network assets;
> - promote the development of partnerships, eg with transport operators;
> - provide travel and other network information to customers, especially the travelling public;
> - ensure a consistent approach to managing the network within a route strategy framework.

3.137 The performance of the network in meeting the new objectives will be measured by a series of indicators to be developed by the Agency and published each year in its annual report. Performance will be reported against both economic and environmental indicators.

3.138 To serve road users more effectively, we have asked the Agency to work on proposals for *Regional Traffic Control Centres* (RTCCs) in England, complementing those already established in Wales. In Scotland, progress has already been made through the establishment and continuing development of the Scottish National Network Control Centre in Glasgow.

3.139 The aim of RTCCs is to:

- improve reliability on the network;
- reduce the disruption caused by major incidents;
- provide re-routing advice to minimise the effect of congestion and incidents;
- minimise delays due to roadworks;
- influence pre-trip decisions on route, time and mode by providing reliable and accurate information.

3.140 RTCCs can help in tackling the effects of traffic congestion by facilitating modern management techniques, including:

- traffic monitoring and modelling;
- strategic traffic control;
- traffic and travel information;
- assistance to the emergency services;
- network performance monitoring and management information.

Helping the road user

3.141 In order to improve the service for transport users, we have asked the Highways Agency to revise the "Road User's Charter" to bring it into line with integrated transport policy and give it an increased customer focus. The Highways Agency will continue to look for greater involvement with users of the network and there will be independent surveys of customer satisfaction.

3.142 Free recovery services at road works have proved successful in removing broken down vehicles quickly and looking after the safety and well-being of drivers and their passengers. The

Highways Agency will build on this experience to improve on response times where breakdowns occur.

3.143 On motorways, following breakdowns or accidents, recovery vehicles are currently mobilised by the police. This service is important in both removing obstructions quickly and securing the safety of drivers and their passengers. The Highways Agency will work with the police to ensure the continuing improvement of this service. **We will also look for ways to give recovery vehicles, which would need to be properly accredited, higher priority in congested traffic, including allowing them to run on the hard shoulder.** These measures to improve the service to motorists will be complemented by our existing programmes to enhance roadside equipment such as CCTV cameras for use by the police and the replacement of older style emergency telephones by those which can be used by disabled people.

3.144 Through the new issue of the Highway Code we will provide clearer advice about the action to take should a motorist breakdown on a motorway. The guidance will explain how to find the nearest emergency roadside telephone. We will look at other ways of making this advice more widely available for motorists, both at the start of and during their journeys and at ways to improve the signing of emergency telephones.

> **Improving roadside facilities for lorry drivers**
>
> - at motorway service areas – we will publish best practice advice for developers and local planning authorities on improving facilities for lorry drivers, including short-stay and overnight parking, toilets and showers, food and refreshments;
>
> - on other trunk roads – updated advice will encourage local authorities to identify locations where roadside facilities are inadequate and to favour proposals that take proper account of the needs of lorry drivers over those that do not;
>
> - through better signing of lorry facilities.

Better information for the driver

3.145 As demonstrated by the AA's *Roadwatch*, the provision of relevant, timely and accurate information can help to make the best use of the road network by enabling travellers to make informed choices about alternative modes, routes and times. The Highways Agency will provide free of charge the roadside information that drivers need to make effective use of their network (as should local highway authorities for their networks). We also want to encourage a competitive market in more specialised travel information services supplied to individuals and companies. We will maintain an appropriate balance between these objectives, exploring the opportunities for public-private sector partnerships to achieve them.

Driver information

pre-trip information

- available in homes and offices from various sources including the radio and internet – including the Highways Agency's website.

in-trip information

- traditional roadside signing and road marking;
- electronic variable message signs;
- in-car radio;
- in-vehicle congestion warning systems;
- route guidance systems.

future possibilities

- Radio-Data System- Traffic Message Channel – pilot service starting shortly;
- dynamic route guidance systems;
- dedicated short range communications (roadside beacons) – three year technical trial (called Road Traffic Advisor) looking at user acceptance and safety.

More care for the local environment

3.146 When we plan trunk roads, we will place greater emphasis on the need to avoid sensitive sites. **Our strong presumption against transport infrastructure affecting environmentally sensitive areas and sites is explained in Chapter 4.** In operating the network, the effects on the natural and built environment will be assessed and where practicable mitigated. For example, we are publishing new advice on reducing the impact of roads on vulnerable species such as otters and bats.

3.147 The Highways Agency has research in hand on various matters, including a joint project with the Environment Agency on the polluting effects of water running off roads. It is also helping to develop new European standards to encourage greater use of recycled materials in construction. More information on the Highways Agency's environmental work will be published later this year.

3.148 Road lighting is needed on some roads in the interests of safety. Where lighting is essential it should be designed in such a way that nuisance is reduced and the effect on the night sky in the countryside minimised.

New lighting for the M62

- installed by the Highways Agency to reduce intrusion into the night landscape where the motorway crosses the high Pennines over Saddleworth Moor;
- new lamps direct most of the light downwards onto the motorway, produce a more natural colour and bring about a dramatic improvement in the night sky;
- lamps are about 30% brighter and last half as long again as those used previously.

3.149 Advice on the design of road lighting[14] has recently been reviewed and expanded to provide up-to-date guidance on the appearance of lighting both during the day and at night. Guidance is also being developed on the assessment of new and replacement lighting schemes.

---

14 in the "Design Manual for Roads and Bridges" volumes 10-11.

CHAPTER 3 Integrated Transport

More care for the environment: cutting down on light pollution.

Better development control

3.150 When responding to development proposals near trunk roads, the Highways Agency will reflect the context established by Regional Planning Guidance and development plans. The Agency will work with local authorities and public transport operators to explore transport options that are sustainable, including those that can be achieved through the use of planning conditions and planning obligations. Regional sustainable transport strategies and *local transport plans* will in due course provide more comprehensive information to support development control decisions (see Chapter 4).

3.151 Previously, the formation of new accesses to trunk roads has been discouraged in order to allow the free-flow of traffic. In support of our integrated transport policy the Highways Agency will in future adopt a graduated policy on new connections to trunk roads. Access will be most severely restricted in the case of motorways and core national routes. Elsewhere, there will be a less restrictive approach to connections, subject to consultation with the local authorities concerned.

3.152 This graduated policy will be of particular value in urban areas where there are brownfield sites that we would wish to see developed in support of our policies for sustainable development. Where brownfield sites could be connected to the trunk road network we will expect proposals for development to support the use of public transport, cycling and walking.

3.153 The Highways Agency will retain the right, on behalf of the Secretary of State, to direct the refusal of planning applications where the proposals raise significant concerns for road safety. Details of the new policy will be provided in a revision to planning policy guidance on transport (see Chapter 4) and an update of the Department of Transport Circular 4/88.

Delivering the goods: sustainable distribution

3.154 We sometimes take for granted how much our standard of living depends on goods delivered by the transport system. The question we face is how to deliver goods efficiently and with least harm to the environment and our health.

3.155 To achieve our aims, we will work in partnership with industry to promote sustainable distribution. By this we mean improving the efficiency of the distribution market in a way that meets our environmental objectives. It also means better planning and higher standards in the industry. We will publish shortly a strategy setting out a wide range of initiatives to deliver these objectives.

Improving efficiency

3.156 Vehicles running empty or lightly loaded lose the industry money, increase pollution and energy consumption and produce unnecessary pressures on road space.

3.157 The proportion of empty running lorries remains significant, at around 30%, and has been broadly static for the last ten years. It is more difficult to ascertain the extent of light running, where lorries are loaded to below their full capacity, but it is substantial. Whilst there are some industries where it is impractical to secure return loads, there are areas where it is possible to reduce light or empty running; for example, through improved information systems and promoting collaboration between operators to consolidate loads into fewer vehicles.

> **Good practice from Tesco**
>
> After completing their deliveries to stores, Tesco's lorries go on to suppliers and collect loads to take back to the distribution centre. Benefits over a full year are three million fewer miles, saving 4,600 tonnes of CO_2 and £720,000 in fuel.

3.158 Fuel efficiency of lorries has improved by some 60% over the past 25 years. Trials by vehicle manufacturers demonstrate that further energy savings could be made by changes to driver behaviour.

3.159 We shall support industry's efforts to realise efficiency gains which deliver wider benefits; for example, through research and benchmarking to identify opportunities for reducing empty and light running, whether through investment in new technology (such as double-deck trailers or IT tools which facilitate load sharing and better route planning) or improving driver training.

3.160 From 1 January next year, we are obliged to conform with EU law by raising the maximum axle weight for lorries on international journeys from 10.5 to 11.5 tonnes and increasing the maximum gross weight of 5 axle articulated lorries from 38 to 40 tonnes. It would be very difficult in practice to distinguish national from international journeys in a way which is both fair and efficient, so we will allow such vehicles for both domestic and international journeys on UK roads. We will bring forward the necessary legal changes shortly.

3.161 These changes will not alter the size of vehicles but will allow more load per vehicle to be carried: this will improve the efficiency and competitiveness of UK hauliers. The problem is that the increased axle loading will cause greater road and bridge wear. A 40 tonne, 5 axle lorry with an 11.5 tonne axle weight causes about a third more wear than the heaviest lorries now permitted for general use (ie 38 tonne vehicle with an axle weight of 10.5 tonnes). Road maintenance is a substantial burden on the taxpayer and it is important that we do all we can to minimise the damage caused by heavier axle weights.

3.162 We are therefore developing a strategy to provide hauliers with incentives to make greater use of 6 axle lorries instead of 5 axle ones. 6 axle lorries are less damaging to roads and bridges because the extra axle allows the weight to be spread more evenly. But the load they can carry is less because the extra axle weighs about a tonne and the lorries are more expensive, making them less attractive to hauliers. The review of the basis of lorry Vehicle Excise Duty rates (VED) already announced by the Chancellor (see Chapter 4) will form part of the strategy by ensuring that the environmental damage, including to roads, caused by different types of lorries is reflected in their VED rates.

3.163 In addition, we want to provide a practical answer to the impact of the extra axle on the load that can be carried. **We have therefore decided to allow 41 tonne gross weight lorries, on 6 axles and with road friendly suspension, on UK roads from 1 January 1999**[15]. These lorries will have to

---

15 Allowing hauliers to operate at 41 tonnes on 6 axles means they can carry approximately the same load as 40 tonne lorries on 5 axles, and still cause considerably less road and bridge wear. This is because their maximum axle weight will be limited to 10.5 tonnes under UK regulations for both domestic and international journeys.

meet the same requirements as 38 and 40 tonne lorries for braking, noise and pollution.

3.164 We have also considered whether to go further and allow for general use the 44 tonne 6 axle lorry which was recommended by Sir Arthur Armitage in 1980[16], and which has been used for combined road/rail transport in the UK since 1994. 44 tonne lorries are effectively the same lorries as existing 38 tonne lorries: they are the same size, they meet the same minimum braking requirements, and the same maximum noise requirements, and their effects on road wear are similar. They would make road haulage more efficient because each lorry can be more fully laden, requiring fewer journeys for the same distribution tasks. Although a 44 tonne lorry would burn slightly more fuel and thus pollute slightly more than a 40 or 41 tonne lorry, the reduction in the total number of lorries for any given amount of goods distributed would bring less pollution overall. Similarly, there would, overall, be less noise, congestion and nuisance, greater safety and less damage to roads and bridges.

3.165 However, a significant disadvantage of allowing 44 tonne lorries for general use is the risk that this could, in some situations, provide an incentive to switch freight from rail to road. One of the key objectives of the *New Deal for transport* is to encourage rail freight as a way of reducing pollution and congestion. Rail freight has benefited from the existing weight concession for combined road/rail movements. While much of the traffic that would take advantage of 44 tonne lorries, such as fuel deliveries to filling stations, is unsuitable for transfer to rail, it seems likely that some existing or future rail freight would transfer to road if 44 tonne lorries were allowed for general use.

3.166 Estimates of the impact of increasing lorry weights on lorry traffic are very sensitive to the assumptions made about the impact on rail freight and how much new lorry mileage would result. It is estimated[17] that if 44 tonne lorries were available now, between 3,000 and 5,000 lorries might be taken off our roads but other than in the short term the numbers of heavy lorries would continue to grow.

3.167 As noted above, we are reviewing the basis of lorry VED rates. In addition, we are bringing forward a number of measures to promote rail freight, and to support the efforts which the freight train operators are now making to turn the tide of 40 years' decline. But it will take time for the full benefits to be realised and we believe it is important to give industry a realistic and increasingly attractive alternative to road haulage. An immediate move to 44 tonne lorries could prejudice that objective.

3.168 **We therefore intend to ask the *Commission for Integrated Transport* (see Chapter 4) to consider the case for allowing 44 tonne lorries, on 6 axles, for general use in the light of the results of the review of the basis of lorry VED rates and evidence from interested parties including the rail freight operators and industry generally.** In bringing forward its recommendations, we will ask the Commission to consider the best solution consistent with our approach for integrated and sustainable transport; in particular, whether there are measures that could be adopted to mitigate the potential impact on rail freight, including phasing of the introduction of 44 tonne lorries to allow more time for rail operators to expand their markets. We will also ask the Commission to consider whether there is scope for limiting any extension to 44 tonnes to lorries with the highest standards of emissions. We would not envisage the implementation of 44 tonne lorries before 2003. It is our intention to give railways the chance to develop the heavy load market.

---

16 "Report of the Inquiry into Lorries, People and the Environment", HMSO, 1980.

17 based on 1996 road and rail freight traffic, current VED rates and on the 5 tonne payload differential between 38 tonne lorries on 5 axles and 44 tonne lorries on 6 axles.

Delivering the goods: sustainable distribution

3.169 Further discussion of the lorry weights issue, along with our detailed proposals for measures to improve the efficiency of lorries and to mitigate their effects on the community and on the environment, will be set out in our forthcoming paper on sustainable distribution.

Quality Partnerships for freight

3.170 We will promote the development of *Quality Partnerships for freight* between the road haulage industry, local authorities and business. The aim will be to develop understanding of distribution issues and problems at the local level and to promote constructive solutions which reconcile the need for access for goods and services with local environmental and social concerns. This will build on existing experience such as 'Delivering the Goods', a joint initiative on urban distribution by the Local Government Association and the Freight Transport Association.

3.171 In our towns and cities, measures aimed at shifting lorry traffic away from the morning and afternoon peak hours could help to alleviate congestion and make better use of local networks. But it is also essential to minimise and avoid increasing disturbance to residents through out-of-hours deliveries.

3.172 The Traffic Commissioners play a central role in the regulation of lorries through their oversight and enforcement of the operator licensing system, which ensures that vehicles are safe and properly maintained, and that operators are fit and proper people to carry out their business. In exercising this role, the Traffic Commissioners' knowledge of the heavy goods vehicle (HGV) industry, their independence and their regional base, are particular strengths that we wish to retain and build on.

Suitable traffic for suitable roads

3.173 The efficient distribution of goods and services must be weighed against concerns about the quality of the urban and rural environment for the people who live and work there. There is substantial concern about the problem of 'rat-running' by large lorries, especially in rural communities. We agree with these concerns. **Lorries should not travel on unsuitable roads unless they have to use them for collection or delivery. There is an established network of primary routes which lorry routeing should follow.**

3.174 We will work with the Freight Transport Association and the Road Haulage Association to develop and publicise their 'Well Driven' scheme which is currently being extended to vans. This scheme provides a mechanism for people to complain about insensitive or irresponsible behaviour by lorry and van operators and drivers, including rat-running on unsuitable roads.

3.175 Bringing forward strategies to keep lorries away from unsuitable areas will be critical issues for local authorities in preparing their *local transport plans*. Under the Road Traffic Regulation Act 1984, local authorities already have powers to prohibit or restrict lorry access, but in certain cases may need the approval of the Secretary of State. We will look at ways of improving and streamlining these arrangements.

3.176 There may also be scope for reducing the number of lorry and van movements by promoting greater consolidation of loads and drawing on the experience of 'City Logistics' systems[18] where goods destined for city centres are diverted into common transhipment facilities with local distribution being carried out using specialised vehicles which may be

---

18 for example, in Germany, the Netherlands, Denmark and Switzerland.

smaller, quieter and less polluting. We will learn from the experiences gained in Europe from operating such systems.

3.177 Where environmental and noise concerns have led to lorry restrictions some firms have already responded with the use of alternatively powered vehicles.

> **More environmentally friendly lorries**
>
> - BOC has vehicles powered by liquefied natural gas. They produce fewer emissions than diesels and are considerably quieter. The project benefits from a 50% grant from Energy Saving Trust's Powershift initiative;
> - Marks and Spencer use quieter, gas powered vehicles to deliver at night in Kensington and Chelsea as an exception to a night-time ban;
> - J Sainsbury is experimenting with a solar powered refrigeration unit, replacing diesel power to cut noise and pollution.

Sustainable air freight

3.178 The increasing demand for rapid distribution of goods will continue to put pressures on air freight services and in turn on airports and associated infrastructure, adding to the pressures from growth in passenger traffic. The rapid growth of air cargo services and their wider economic, environmental and social significance requires further examination. We will commission new research to inform future policies on the air freight industry. The research will:

- assess the current development of the sector, including its economic importance and wider impacts;
- provide a better basis for forecasts of its future growth and the implications for demand for services and market change;

- support the development of the new national airports policy, which will set the framework within which the industry can plan for the future with greater certainty.

Sustainable shipping

3.179 The decline in the British merchant navy was accepted by the previous Government as the inevitable outcome of market forces. But the international market is significantly distorted by the effects of cut-price shipping and foreign subsidies.

3.180 We will take a strategic view of the role of shipping and the wider maritime-related industries in the national economy so as to determine Britain's future maritime needs and how those may be secured. **Our policy will be based on a broader, long-term vision of the importance of British shipping to the nation**. We will establish a clear set of objectives with firm commitments to action agreed jointly by the industry, unions and Government.

3.181 This integrated shipping policy will have four broad aims:

- to facilitate shipping as an efficient and environmentally friendly means of carrying our trade;
- to foster the growth of an efficient UK-owned merchant fleet;
- to promote the employment and training of UK seafarers in order to keep open a wide range of job opportunities for young people and to maintain the supply of skills and experience vital to the economy;
- to encourage UK ship registration, to increase ship owners' identification with the UK, to improve our regulatory control of shipping using UK ports and waters and to maintain the availability of assets and personnel that may be needed in time of war.

Delivering the goods: sustainable distribution

We shall encourage UK ship registration: cargo vessel at Tilbury docks.

3.182 We are committed to working with the shipping industry to develop its potential to the full. We set up a Shipping Working Group last year to consider how to obtain the maximum national economic and environmental benefit from shipping. The Group reported in March with a range of proposals on seafarer training, employment, the fiscal environment and opportunities for UK shipping. **Our response to these proposals and our strategy for reviving the shipping industry will be published shortly**.

Making better use of coastal shipping and inland waterways

3.183 Research[19] has indicated that there may be potential to divert about 3.5% of the UK's road freight traffic to water, split between ships re-routing to ports nearer to the origin and destination of their loads and the potential for bulk and unit loads to shift to coastal traffic.

3.184 We intend to bring forward legislation to extend the application of the freight grant regime to include coastal and short sea shipping, reflecting a recommendation of the Shipping Working Group. We will consult on the details, including the costs which would be eligible for grant and the criteria to be used in assessing applications.

3.185 We will also encourage greater use of inland waterways, where that is a practical and economic option. We will re-examine the rules of the freight grant regime with a view to encouraging more applications for inland waterways projects. We want to see the best use made of inland waterways for transporting freight, to keep unnecessary lorries off our roads.

3.186 In addition to carrying freight, inland waterways also have an important role to play in providing leisure and tourism opportunities and can provide a catalyst for urban and rural regeneration.

3.187 Our revised planning guidance will encourage more freight to be carried by water. Local authorities in their development plans will be expected to consider opportunities for new development which are served by waterways.

Moving more goods by inland waterways: freight on its way to Leeds.

---

19 "Roads to Water Research Project", Jonathon Packer and Associates, 1993.

CHAPTER 3 — Integrated Transport

> **Thames 2000**
>
> Aims to establish new passenger services on the River Thames for the Millennium Exhibition at Greenwich and leave a lasting legacy of improved infrastructure and services. It has three key aspects:
>
> - new passenger services, announced in March, including links to the Millennium Experience
>   - express services from dedicated London piers;
>   - shuttle service linking Greenwich town with the Millennium site;
>
>   ... and longer term legacy services
>   - a 'hopper' service linking key central London destinations;
>   - an express service to central London;
>
> - a programme of infrastructure works to create up to ten new piers at key locations on the river, modernise existing piers and improve linkages with other public transport;
>
> - a new London Transport subsidiary – London River Services Ltd – to own and manage key piers on the river and promote, license and co-ordinate passenger services on the Thames, to help ensure that river services are integrated into transport plans for the capital.

3.188 The River Thames is a greatly under-used asset in London. It has potential for passenger transport and for freight, including aggregates and the transfer of waste. We are working to unlock this potential, through Strategic Planning Guidance for the Thames and through our Thames 2000 initiative which will establish new passenger river services by the Millennium. We will also ensure that use of the river is more fully integrated with other transport services in London, especially bus services.

Better integration of airports and ports

Integrated airports

AIRPORTS POLICY

3.189 As recommended by the Transport Select Committee in May 1996[20], **we will prepare a UK airports policy looking some 30 years ahead**. This will develop the application to UK airports of the policies set out in this White Paper – of sustainable development, integration with surface transport and contribution to regional growth.

3.190 It will provide the framework within which those concerned can plan for the future with greater certainty. We will consult widely in preparing the new policy and will take account of the Inspector's report on the Heathrow Terminal 5 inquiry.

3.191 The policies we bring forward for civil aviation, as for other forms of transport, will reflect our strategy for sustainable development. This means aviation should meet the external costs, including environmental costs, which it imposes. We must tackle the effect of civil aviation and airports on the environment (see Chapter 4).

3.192 The new airports policy will take account of the demand for airport capacity for scheduled, charter, business and freight aviation and the related environmental, development, social and economic factors. It will be taken forward in conjunction with airspace capacity issues and with consideration of surface access provision, particularly better public transport access. It will also consider ways, whether by economic or

---

20 Second Report of the Transport Committee, House of Commons Session 1995-96, on UK Airport Capacity, published 21 May 1996, HoC paper 67.

regulatory measures, of improving the utilisation of existing capacity, where this might be desirable; and it will take into account possible future developments in European legislation, for instance on runway slot allocation and airport charging.

Less congested airports can relieve pressure on major airports.

3.193 The new policy will reflect the different roles and competitive strengths of the nation's airports. The largest and busiest airports serve the whole country or a large part of the country, and offer frequent direct services to a wide range of destinations. Many airports thrive on serving a more local area, with a combination of direct services where the demand is sufficient and connections to major international hub airports. Less congested airports, such as Luton, Stansted and most regional airports, can also be attractive to the new generation of low-cost airlines.

3.194 Each airport cannot be viewed in isolation from other airports. Airports both compete with each other and complement each other to some extent. A good example of this is Manchester and Liverpool: we welcome the co-operation between these airports which has developed during the last year. The new airports strategy will consider how each region might best be served by the combination of the available airports in the region; and, how regions and their airports, for example in the North and Midlands, might work together to realise the potential of airports away from the congested south east of England.

3.195 The policy will draw on new studies of the role of airports in economic development to gain a better understanding of the underlying relationships. These studies will focus on both aviation opportunities and the link between air services, economic growth and regeneration in specific local circumstances.

Role of Regional Airports

- SW England study – underway, expected to report by late 1998;
- studies to be carried out in Scotland, Wales, Northern Ireland, the Midlands and the North of England: phased programme starting summer 1998 and reporting in 1999;
- studies will be carried out in close consultation with local authority representatives and other interested organisations, including Regional Planning Conferences.

3.196 Taking account of the emerging findings of these studies, we will encourage the growth of regional airports to meet local demand for air travel where consistent with sustainable development principles. The aim is to:

- maximise the contribution which they make to local and regional economies;
- relieve pressure on congested airports in the south east of England;
- reduce the need for long surface journeys (particularly by road) to south east airports.

3.197 **We have recently announced proposals to encourage international flights to regional airports through a policy of greater liberalisation.** We have decided that open access to all UK airports, except Heathrow and Gatwick, should be offered to all of our bilateral air service partners, provided that UK airlines are also allowed to operate on the same routes. This change will allow both UK airlines and airlines of the country concerned to operate to and from that country on such routes without restrictions on capacity or frequency, and without the need for international aviation negotiations to establish such services. This will enable UK and foreign airlines to plan the development of services with confidence that future growth will not be limited by bilateral restrictions.

3.198 We have also announced proposals to free soundly-financed local authority airports from public sector borrowing controls. This relaxation will greatly assist major regional airports to invest and expand when this is commercially justified. It will allow them to compete for business on a level playing field with private sector airports.

3.199 We will also press for recognition in the revised EU regime for slot allocation, of the case for maintaining access from the regional airports into major hubs such as Heathrow and Gatwick.

AIRPORTS AS INTERCHANGES

3.200 **Airports are key interchanges** and major employers. Improving access to them by public transport will help to reduce congestion and pollution on nearby roads.

3.201 Our consultation demonstrated a clear willingness on the part of the aviation industry and other interested parties to tackle the problems of airport accessibility. Airports such as Heathrow and Manchester already have programmes to improve public transport access. These are designed to get passengers and employees to use their cars less by improving the public transport alternatives.

3.202 As managers of some of the nation's largest public transport interchanges, airport operators will be well placed to make a positive contribution to integration. We will therefore expect airport operators to be partners in implementing surface transport initiatives to improve the quality of the public transport journey to their airports. The support of airlines using the airport is also important.

Manchester Airport

The airport's 'regional transport strategy' has a vision of integrated transport based on partnership to:

- increase public transport use by passengers and staff from 10% in 1992 to 25% by 2005;
- develop high quality ground transport interchange – construction of the first phase, a bus and coach station, begins later this year;
- improve the airport's rail connections – building on the frequent direct rail services to many major towns and cities in the North and Midlands;
- develop and promote a green commuter plan to increase environmental awareness among its employees.

3.203 The needs of surface access to airports should be considered as part of the wider transport strategy for the local area. Airport-related transport issues must be integrated with, not divorced from, local transport problems and opportunities.

LOCAL CONNECTIONS

3.204 *Local transport plans* should reflect the wider transport role defined for airports in regional strategies. To complement this work, we consider that all airports in England with scheduled passenger services should lead an *Airport Transport Forum*. Some of the larger airports have found these groups valuable in ensuring co-operation between all those interested in the development of surface transport serving the airport.

> **Funding local improvements**
>
> - some measures require only minimal funding (eg shared taxi schemes). Other improvements can flow from modest start-up funding (eg new bus routes or park and ride schemes). Some proposals, even those which would be commercially viable in the longer term, may require substantial development finance (eg new rail links);
>
> - possible sources for funding include:
>
>   - from the aviation industry – where a scheme is viable or there are wider benefits to the industry;
>
>   - for airports to levy a surcharge on car parking charges;
>
> - with both options we would expect the proceeds to be applied to public transport improvements or measures to mitigate the undesirable impacts of road traffic to and from the airport.

3.205 We envisage that local authorities, including the Passenger Transport Authority where applicable, would participate in the Airport Transport Forum which should have three specific objectives:

- to draw up and agree challenging short and long term targets for increasing the proportion of journeys to the airport made by public transport;

- to devise a strategy for achieving those targets, drawing on the best practice available. This is likely to involve a wide range of measures to address the needs of all those travelling to airports. Bus and coach services should be included as well as rail. This means that the management of traffic on local and trunk roads will be an important issue for some airports. We would hope to see strategies agreed by late 1999 and fed into the development of *local transport plans*;

- to oversee implementation of the strategy. Implementation should include green transport plans to cover commuting and business travel for all employees based at airports.

NATIONAL CONNECTIONS

3.206 Integrating airports into the wider transport networks also means developing the connections to national and regional rail and coach services to reduce the present reliance on private, road based transport. While the new core national route network in England recognises the importance of airport connections, **we will be looking for opportunities to facilitate public transport links to airports, with a particular focus on improved rail access.**

3.207 Improving access to the UK's major hub airports by rail from other regions has the potential to attract feeder traffic away from roads (or even air) and bring environmental benefits. The *Strategic Rail Authority* will consider rail schemes that address deficiencies in direct airport links to the national network and encourage the development of regional and long-distance feeder services. We will make improving rail access to airports one of its aims.

3.208 Major new rail infrastructure is expensive. Links to thriving airports will have to compete against other claims on Government expenditure. We would expect the aviation industry itself to contribute funding for improvements, taking account of the extent that it benefits.

BAA working to integrate airports

- linking Heathrow to the national rail network – £440 million investment in the Heathrow Express to improve links with the national rail network and increase the proportion of the airport's passengers on public transport from a third to BAA's target of 50%;

- attracting airport staff on to buses – BAA has increased the quality, frequency and reliability of Heathrow and Gatwick local bus services to persuade the airport's staff to leave their cars at home. A pump-priming strategy has led to nearly a threefold increase in passengers;

- Stansted rail links – a local service between London and Stansted has recently been introduced by West Anglia Great Northern to complement the SkyTrain express service. A service to Stansted from Birmingham has resulted from partnership discussions between BAA, Essex and Cambridgeshire County Councils and Central Trains Ltd;

- the Heathrow Area Transport Forum – a forum of local transport authorities, key local businesses and transport operators co-ordinating transport policy across the area;

- Heathrow travelcard – an innovative travel card which entitles 56,000 staff working at Heathrow to discounts of up to 80% on 17 bus and coach services.

Integrated ports

3.209 Ports are a vital link in the supply chain to and from our trading partners and must be integrated with wider transport networks. The aims of our policy will be to:

- promote UK and regional competitiveness by encouraging reliable and efficient distribution and access to markets;

- enhance environmental and operational performance by encouraging the provision of multi-modal access to markets;

- make the best use of existing infrastructure, in preference to expansion wherever practicable;

- promote best environmental standards in the design and operation of ports, including where new development is justified.

3.210 The *Strategic Rail Authority* will be responsible for reviewing the scope for improving rail access to major ports, in consultation with Railtrack, the rail freight industry, port owners and shippers. Some improvements have already been made or planned – for example, Railtrack has started work on increasing capacity on the routes to Southampton and Felixstowe. The *Strategic Rail Authority* will need to consider whether further improvements are feasible and, if they are not commercially viable, whether it is justified to give some support from its own budget.

3.211 As port expansion can have significant effects on sensitive marine environments we will encourage the ports industry to invest in measures to deal with increased demand whilst avoiding the physical expansion of port land. English Nature is developing best practice on coastal management which will cover the role of ports.

3.212 The European Commission has recently published a Green Paper on ports and maritime infrastructure[21] which states that the main objectives for ports should be to increase their efficiency and improve infrastructure by integrating ports into the multi-modal Trans-European Networks (TENs) and to ensure free and fair

---

21 "Sea Ports and Maritime Infrastructure", European Commission Green Paper, COM (97)678, 1997.

competition. We strongly support the Commission's proposals, recognising the importance of environmental issues in port development.

Trans-European Networks

3.213 We will continue to work with the EU on the development of TENs. Our approach will be to seek to ensure that funding is directed at proposals which demonstrably further both European and UK transport objectives; and in particular shift passengers and freight from road to rail.

3.214 We will continue to make the best use of European funding of TENs in support of projects that help to improve strategic transport links between the UK and the rest of Europe. For at least the duration of this Parliament, we will continue to bid for support for the UK's two high-speed rail priority projects – the West Coast Main Line modernisation and the Channel Tunnel Rail Link. In addition, opportunities will be explored for gaining support from the TENs budget towards other worthwhile projects that support our integrated transport policy.

Improving rail connections to the rest of Europe.

Channel Tunnel Rail Link

Public-private partnerships are back on track with the revised agreement to build the Channel Tunnel Rail Link and operate the Eurostar service. £1.8 billion of Government grant will be complemented by £3.7 billion of private funding, which in a unique ground-breaking development will be raised through Government backed bonds. The link will be built in two phases and should be completed by 2007. The deal will deliver:

- a dedicated high speed railway for Channel Tunnel traffic, providing a strategic economic artery for international and domestic passengers and freight;

- a new international and domestic multi-modal transport interchange centred on Heathrow airport, providing an international gateway for rail services across the UK;

- over £3 billion of economic, transport and environment benefits including a major boost to the economic regeneration of North Kent and the Thames Gateway.

The Channel Tunnel Rail Link is a tangible example of our commitment to integrate national and international transport systems.

3.215 Reform of the railways across Europe is essential if rail is to deliver seamless and sustainable trans-European services capable of serving the needs of the Single Market. We welcome and will continue to press for progress towards the implementation of the recommendations set out in the European Commission's 1996 White Paper "A Strategy for Revitalising the Community's Railways".

3.216 Last November we secured important rail freight commitments from Eurotunnel and the French Government as part of the price for agreeing to an extension of Eurotunnel's concession. The deal includes agreement by the French Government to establish rail freight corridors to give international freight a higher priority. This package will help to realise the full potential of the Channel Tunnel for long distance rail freight.

Travelling safely

Road safety

3.217 The numbers killed on our roads are equivalent to 30 average commercial[22] aircraft, fully loaded, crashing in the UK every year. But because road casualties occur only a few at a time they are not always noticed as much as aircraft or train disasters where, overall, the number of people killed is very much lower.

3.218 In 1987 a target was set to reduce road traffic casualties by a third by the year 2000 compared to the average for 1981-85 and this had a major influence in raising the profile of road safety. By 1997 the number of deaths on the road had fallen by 36% to 3,599 and the number of serious casualties had declined by 42% to 42,967. The total number of casualties has, however, not gone down, standing at 327,544.

3.219 In our Manifesto, we said that we would make road safety a high priority, that cycling and walking must be made safer especially around schools. **As part of the *New Deal for transport* we will set a new road safety target for Great Britain for 2010[23] which we shall publish later this year.**
We will at the same time set out a strategy and programme of measures for achieving it.

3.220 To improve road safety and save lives, action must be taken across a number of fronts – including improvements in the behaviour of drivers, riders and pedestrians; enhancements in vehicle safety; better roads and road engineering; and better enforcement. It will also require the positive co-operation of many organisations, including local authorities; the police; schools; the motor manufacturers; and indeed all road users themselves and their associations. One of the main elements of the strategy will be to involve all these agencies in the achievement of the new target. We do not want to make roads safer by simply discouraging vulnerable groups from venturing on to roads.

Improving road safety

- reviewing the driving test and driver training, to develop a more effective test and better training techniques;

- improving road safety education in schools and by parents, by assessing the effectiveness of existing training aids and developing new ones;

- assessing local measures to achieve safer routes to school, and producing a best practice guide;

- surveying potential measures to ensure better compliance with speed limits in urban areas and on rural roads – the two most dangerous types of road;

- research into measures to improve vehicle safety and to ensure that they give maximum protection to occupants and minimise injury to pedestrians and cyclists.

---

22 Boeing 737 taken as an example.

23 The Department of the Environment, Northern Ireland is carrying out a similar exercise.

3.221 We wish particularly to improve the safety of more vulnerable road users, including pedestrians (particularly children), cyclists and motorcyclists, in a way that is consistent with encouraging more cycling and walking. We want our children to be able to walk to school in safety: initiatives supporting safer routes to school will support both safety and environmental aims.

3.222 We will look at how to improve the safety of novice drivers – who are involved in nearly a fifth of the total number of casualty accidents – and at measures to reduce speed related accidents. Speed is thought to be a factor in about a third of all casualty accidents. In partnership with industry, we will encourage better driving by professional drivers – both lorry and bus drivers and those driving company cars on business (who are disproportionately involved in accidents).

3.223 Drink-driving is still a major cause of deaths, and we have consulted recently on proposals for a package of measures to combat this continuing problem[24]. These included possible measures to improve enforcement and education and we sought views on whether the current legal blood alcohol limit of some 80mg per 100ml should be reduced to 50mg. In the light of the responses to the consultation, we hope to announce our conclusions later this year.

3.224 Another major theme will be the scope for improving safety through the better enforcement of existing regulations.

3.225 Measures to improve road safety will also contribute to the efforts towards the proposed target in our Green Paper "Our Healthier Nation", to reduce the number of major accidents from all causes by one-fifth by 2010. All the relevant Government departments are collaborating to ensure consistency of approach on this.

3.226 The EU also has a role in promoting road safety. Amongst other things, it plays an important role in establishing technical standards for vehicles and has set up a number of working groups which have produced proposals on further measures to improve safety. We will continue to work with the EU on road safety initiatives, in particular on the development of higher vehicle safety standards, including those which minimise the impact of collisions on vulnerable road users.

Working within the European Union

- supporting the FIA's "10 seconds which could save your life" campaign aimed at seat belt wearing and other safety measures;

- looking with the European Commission at EU-wide regulation of car advertising on TV using the UK's voluntary code of practice as a model. The code requires that adverts

    - should not encourage or condone dangerous, inconsistent or competitive driving practices or breaches of the Highway Code

    - should not portray speed in ways which might encourage motorists to drive irresponsibly or break the law

    - should not include references to power or acceleration implying that speed limits may be exceeded and there must be no accompanying suggestion of excitement or aggression;

- pressing for EU regulation to make car fronts less dangerous for pedestrians involved in an accident;

- actively supporting the European New Car Assessment Programme (Euro-NCAP) that provides comparative consumer information about the relative crashworthiness offered by new cars on sale to the public;

- saving lives by introducing front underrun guards on lorries – we will consult later this year on the introduction of regulations.

---

24 "Combating Drink-Driving – the next steps", DETR, February 1998.

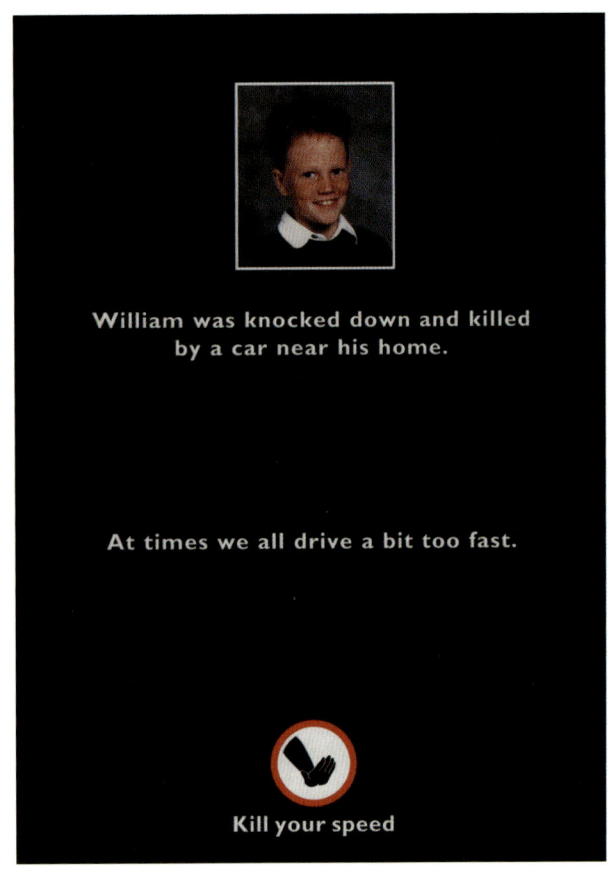

William was knocked down and killed by a car near his home.

At times we all drive a bit too fast.

Kill your speed

Media campaigns put across the message to slow down.

REVIEW OF SPEED POLICY

3.227 Many measures that would help the achievement of the new road safety targets will bring wider benefits for integrated transport policy. Better enforcement of speed limits on all roads would reduce the number of accidents and their severity (see Chapter 4). Lower speeds combined with a more fuel efficient driving style could also bring environmental and social gains and in some circumstances could contribute to the more efficient use of roads in congested conditions.

3.228 But the precise balance between speed reduction for road safety, for social gains and for reducing vehicle emissions, including noise, is not fully understood. Many of the responses to the consultation suggested that we should review speed policy. **We will therefore set up a review to develop a speed policy that takes account of the contribution of reduced speeds to environmental and social objectives as well as to road safety.** We will consider issues such as driver attitudes and how behaviour can be improved through education and enforcement.

3.229 The review will examine how existing best practice in engineering, enforcement, education and publicity can be developed. The aim will be to develop a practical and cost-effective approach which meets our wider policy objectives.

3.230 The review will cover all types of road in Britain, both in town and country. We expect the review to take about one year to complete. We will consult widely, including those environmental groups who traditionally have not been involved in road safety matters.

Speed kills

- Speed – we will continue to campaign in the media to get across the rapidly increasing likelihood of serious accidents as speed increases;

- '20 mph zones' – we will continue to help local authorities fund traffic calming measures and make it easier to introduce 20 mph speed limits;

- speed and red light cameras – we are looking at the funding of cameras and their operation;

- cameras at road works – we will step up the practice of placing speed cameras at road works on motorways and trunk roads.

Travelling safely

Road safety education: cycle training.

MOTORCYCLING SAFETY

3.231 Despite the real and very welcome reduction in the number of motorcycling casualties in recent years (although last year reversed the trend) there were still over 24,000 motorcycle riders and their passengers killed or injured on roads in 1997 – 7.5% of all casualties but 14% of deaths and serious injuries. In built-up areas, motorcycles are three times more likely than a car to have an accident involving a pedestrian.

3.232 One of the concerns raised by motorcycle groups is that the high casualty rate of motorcyclists is due to vehicle drivers not taking enough account of their needs. We have therefore introduced more questions in the driving theory test to increase awareness of vulnerable road users, including motorcyclists. We are also considering what, if any, improvements need to be made to the practical car driving test.

3.233 Training has played an important part in reducing the number of casualties and we will consider how the road skills of riders can be further improved in the future. We will issue a consultation paper soon inviting views on the period of validity of provisional motorcycle licences.

BUS AND COACH SAFETY

3.234 Buses and coaches have a good safety record. The operator licensing system, administered by the Traffic Commissioners, will continue to play a vital role in supervising entry to the bus industry and ensuring safe operation. The Vehicle Inspectorate also has an important role in enforcing road worthiness standards. These controls will remain and their effectiveness kept under review.

3.235 Investment in improving the quality of vehicles and infrastructure which will be encouraged by our policies for the bus industry should bring safety benefits in addition to encouraging public transport use.

3.236 Most bus passenger accidents are the result of falls on the bus or when getting off. But buses in towns are frequently involved in accidents with pedestrians, though the reasons for this are unclear. We plan further research on safety at bus stops. The siting of bus stops and location of crossings should take account of the need to minimise the risk of accidents, whilst encouraging a safer, more pleasant walking environment.

3.237 Other research projects are looking at bus passenger safety, and all casualties in accidents involving buses, coaches and minibuses to see if changes to the construction standards for these vehicles could improve safety for passengers and for other road users. It is now a requirement that seat belts are fitted on some seats in coaches and minibuses, and that a forward facing seat fitted with a seat belt is provided for children in these vehicles when on an organised trip.

3.238 We will also be consulting on changes to **require the fitting of seat belts on all seats in new buses, coaches and minibuses which do not carry standing passengers.** All these measures are designed to make bus travel safer and, thus, to encourage bus use.

DRIVERS' HOURS

3.239 At present there are differences between EU Regulations which govern drivers' hours, (affecting most HGV drivers and around half of the bus and coach drivers in the UK), and UK domestic legislation, (affecting mainly bus and coach drivers and some HGV operations). There is scope for confusion and some difficulty in enforcing the UK legislation which does not require the use of tachographs. The Transport Select Committee has recommended[25] that domestic drivers' hours rules be phased out in favour of the EU rules. **We therefore propose to consult on legislative changes which would bring most operations within the scope of the EU rules.**

3.240 The Commission is currently considering an extension of the Working Time Directive to the transport sector. We support this in principle. There is no reason why transport workers, including professional drivers, should not have the same level of protection against working excessive hours as workers in other sectors. It will, however, be necessary to preserve some flexibility and to take proper account of complicating factors such as the relationship between working time and drivers' hours (for workers covered by drivers' hours regulation), between third party and own account transport operations and between employed and self-employed workers.

Railway safety

3.241 Already rail is one of the safest forms of travel and the long term improvement in rail safety is continuing. But the Chief Inspector of Railways has warned that some operators have tried to avoid taking measures to improve rail safety standards, or worse still, to reduce them[26]. It is vital to ensure that there is no erosion of safety standards in the privatised railway. Existing standards of health and safety must be maintained and, where necessary, improved. **Operators must not put commercial considerations ahead of safety.**

3.242 The Health and Safety Commission (HSC), together with its operational arm the Health and Safety Executive (HSE), which includes the Railway Inspectorate, is the independent regulatory body responsible for railway safety. The Railway Inspectorate has comprehensive powers to enforce the wide-ranging duties of the Health and Safety at Work etc. Act 1974 (HSWA) and associated Regulations.

3.243 A new safety regime for the privatised railway was put in place in 1994. It reflects Railtrack's and train operating companies' obligations under the HSWA to operate safely. Railtrack has responsibilities for both setting and enforcing safety standards. The single most important element in the regime is a requirement for each operator to prepare, and obtain acceptance of, a 'safety case' – a detailed document describing the operator's risk assessments and safety management systems. The regime also gave Railtrack wide-ranging responsibilities for both setting and enforcing safety standards.

---

25 Fifth Report of the Transport Committee, House of Commons Session 1995-96, on the adequacy and enforcement of regulations governing heavy goods vehicles, buses and coaches, HoC Paper 356-I.

26 "Railway Safety, HM Chief Inspector of Railways' Annual Report on the safety record of the railways in Great Britain during 1996/97". ISBN 0-7176-1464-6.

3.244 We are determined to ensure, as part of improving the railways in the interests of passengers, that safety is not compromised.

3.245 The Health and Safety Commission fully shares this resolve. It has recently gone out to formal consultation on draft regulations to oblige the privatised industry to replace or modify Mark 1 (ie slam-door) rolling stock by 2003 and to install train protection (which would apply the brake automatically in danger situations) on all trains and at all key signals by 2004. Mark 1 rolling stock has been criticised because of how it performs in certain types of accidents. The Commission has recommended that all Mark 1 rolling stock be withdrawn by 1 January 2003 unless it has been rebodied by then (in which case it can remain in service indefinitely) or it has been modified to prevent one vehicle overriding another in the event of a crash (in which case the modified stock can remain in service only until 1 January 2007).

3.246 The Commission has brought forward its planned review of Railtrack's role in setting safety standards. As recommended in the recent report from the Environment, Transport and Regional Affairs Select Committee[27], the Commission will be focusing on the functions and responsibilities of Railtrack's Safety and Standards Directorate and whether those functions and responsibilities should remain with Railtrack or should be located elsewhere. In the meantime, the HSE will continue its independent monitoring, investigation and enforcement of railway safety.

Marine safety

3.247 The DETR's newly formed Maritime and Coastguard Agency (MCA) is responsible for developing, promoting and enforcing marine safety standards for the UK and for organising the response to incidents at sea and on the coast, whether they involve danger to life or to the environment. To form the new Agency, we brought together from 1 April 1998 the former Marine Safety and Coastguard agencies.

3.248 Our aim has been to create a single, better integrated, Agency; able to carry out its functions more effectively. For example, the new Agency will be able to use its combined presence around the coast to improve its oversight of leisure craft and fishing vessels, taking an integrated approach to information and education for crews, and to the monitoring and implementation of safety standards; and it will be better placed to enhance the surveillance and control of traffic through the Dover Straits, the busiest sea lane in the world.

> **Coastguards on watch 24 hours a day**
>
> The Maritime and Coastguard Agency will continue to maintain a 24-hour co-ordinating capability for the UK Search and Rescue Region. Designated rescue centres are constantly manned by Coastguard personnel, highly trained in search and rescue procedures.

3.249 **The safety of passengers and crew at sea is vital**. We believe strongly in accident prevention. The new Agency will therefore continue the work of its predecessors in the setting, inspection and enforcement of maritime safety standards. These standards are based primarily on those agreed internationally by the International Maritime Organisation (IMO), strengthened in some cases by regional agreements with our EU partners and other North European countries. The MCA will pay particular attention to the safety of passenger ferries and of bulk carriers and vessels carrying hazardous or polluting cargoes.

---

27 Third Report of the Environment, Transport and Regional Affairs Committee, House of Commons Session 1997-8, on the proposed Strategic Rail Authority and Rail Regulation, HoC paper 286-I, March 1998.

3.250 The MCA will itself enforce safety standards on UK-registered ships. It will also play its full part with our neighbours in enforcing standards on foreign ships through 'port state control'. This will include enforcement of the IMO's International Safety Management (ISM) Code which came into force on 1 July 1998. The ISM Code has broken new ground in seeking to develop a safety culture embracing operations both on ship and ashore. But the only wholly effective way of improving the safety of foreign shipping will be by improving the performance of those flag states, including a number of flags of convenience, who fail to fulfil their commitments under the IMO's Conventions. So the MCA will pursue the campaign of the UK and other like-minded states in the IMO to agree binding and enforceable criteria for the performance by flag states of their obligations.

3.251 We will also look to the MCA to take forward new tasks in relation to ports. Among these will be overseeing effective waste management planning by ports and helping to develop and monitor a new port safety code (see Chapter 4) that we propose to draw up in the light of the official accident report on the grounding of the oil tanker Sea Empress at the entrance to Milford Haven in 1996.

3.252 Effective accident investigation is a key contributor to marine safety. Recent years have seen enormous advances in the technology for locating, exploring and photographing wreckage on the seabed, as was dramatically shown recently by the investigation of the MV Derbyshire that we co-financed with the European Commission, the report on which was published in March 1998.

> **Investigating marine accidents – MV Derbyshire**
>
> - innovative joint DETR/European Commission survey to examine wreck;
> - demonstrated that investigation was possible in extreme conditions;
> - Assessors' report identifies possible changes in vessel design and operation – will help to improve safety;
> - Assessors' recommendations have been put to the International Maritime Organisation.

3.253 The work done on the MV Derbyshire was costly and not affordable for the generality of accidents. But we are considering ways of making cheaper equipment available on a more regular basis to the Marine Accident Investigation Bureau (MAIB) to enhance its accident investigation capability. We will be consulting shipping interests on how they might contribute to the cost of this facility.

Air safety

3.254 The Civil Aviation Authority (CAA) is responsible for the regulation of safety in the UK civil aviation industry and imposes very high safety standards. Consequently, UK airlines have an excellent safety record. In Europe 27 countries, including all EU/EEA Member States, are co-operating to harmonise aviation safety standards to a higher level and there are plans to establish a new European Aviation Safety Authority.

3.255 A significant proportion of passenger services to and from the UK are undertaken by foreign airlines and concern has been expressed about the ability of some foreign countries to ensure that their aircraft meet international safety standards, set by the International Civil Aviation Organisation (ICAO). ICAO has therefore established a Safety Oversight Programme to assess the position and to assist countries that are found to fall short of the required standards. The UK played a leading role in establishing this programme, and fully supports it.

3.256 CAA aviation safety specialists have been made available to participate in ICAO assessment teams and we have provided some direct funding for the programme. In addition, the European Civil Aviation Conference (ECAC) is co-ordinating a Safety Assessment of Foreign Aircraft (SAFA) programme. Under this programme, foreign aircraft which are suspected of non-compliance with ICAO standards are inspected and action is taken to prevent an aircraft from flying if serious problems are identified.

3.257 The Environment, Transport and Regional Affairs Committee's recent report on Air Traffic Control[28] highlighted its concerns about the delays in opening the new National Air Traffic Services (NATS) air traffic control centre at Swanwick. The Committee also recommended that the CAA's Safety Regulation Group should be moved to a new independent transport safety authority, which would include representatives of operators, consumers and those who work in the industry. We have responded to the Committee's recommendations. Our proposals for a public-private partnership for NATS (see Chapter 4) would mean that air safety regulation would be independent of service provision but more generally we intend to review the arrangements for public transport safety (see below).

An integrated approach to transport safety

3.258 DETR with its Maritime and Coastguard Agency, together with the Civil Aviation Authority and the Health and Safety Commission and Executive act as regulating bodies for all transport safety. DETR brings together a range of responsibilities for safety in the transport field, this provides new opportunities to compare and share best practice between the different modes.

3.259 The Environment, Transport and Regional Affairs Select Committee has recommended[29] that transport safety regulation should be focused on a single independent authority, as a means of separating safety regulation from operational responsibilities. This is an important issue with wide reaching implications for current organisational arrangements. **We therefore propose to carry out a review of the arrangements for transport safety, including accident investigation**, with a view to producing a substantive response to the Select Committee's reports in due course.

---

28 Fourth Report of the Environment, Transport and Regional Affairs Committee, House of Commons Session 1997-8, on Air Traffic Control, HoC paper 360-I, March 1998.

29 Third and Fourth Reports of the Environment, Transport and Regional Affairs Committee, House of Commons Session 1997-8, March 1998.

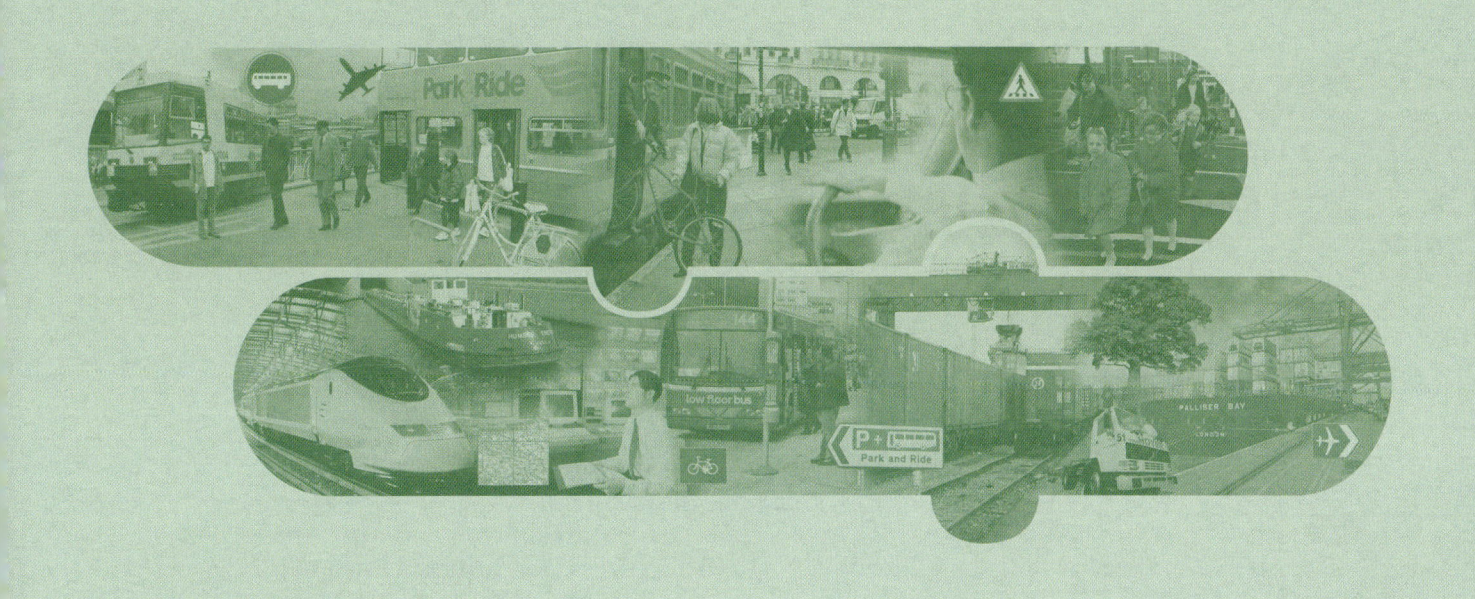

PART 3

CHAPTER 4   Making it Happen
CHAPTER 5   Sharing Responsibility

A New Deal for transport: better for everyone

CHAPTER 4
Making It Happen

> "What counts is what works. The objectives are radical. The means will be modern."
>
> Labour Party Manifesto
> 1997

European action

4.1 The UK cannot succeed in delivering an integrated transport policy in isolation from Europe. The European Union has an important role to play – in setting the framework of policy and law at European level and in promoting partnership and co-operation between Member States, industry and the community. For example, through initiatives such as the European Community strategy for reducing CO_2 emissions from passenger cars, in proposals for revitalising EU railways and a forthcoming paper on infrastructure charging.

4.2 The UK Presidency has provided an excellent opportunity for us to build better, more productive relations with our European neighbours. We will continue to play an active and positive role in future, in particular to help develop European policies which support sustainable transport.

UK action

4.3 The merger of the former Departments of the Environment and Transport has already secured better integration of transport and environmental thinking and land use planning policy. This White Paper sets out the national framework for an integrated transport policy within which others can act.

Commission for Integrated Transport

4.4 This is the first comprehensive White Paper on transport policy for 20 years. But it is not the end of the story: we need to continue to work on our policies and not wait another generation before we take stock of how we are getting on. To help keep the debate alive and to continue building on the consensus, **we will establish a new independent body – the *Commission for Integrated Transport* (CfIT)** – to provide independent advice to Government on the implementation of integrated transport policy, to monitor developments across transport, environment, health and other sectors and to review progress towards meeting our objectives. Its remit will include:

- reviewing and monitoring progress towards objectives and targets set out in the White Paper;
- continuing and refreshing the transport policy debate;
- fostering consensus among practical providers;
- identifying and disseminating examples of best practice from home and abroad;
- advising on developments in Europe, including relevant EU initiatives;
- advising on the role of existing and emerging technologies.

We will ask the *Commission* to advise us, among other things, on:

- setting national road traffic and public transport targets;
- the revisions we will be making to the 1997 National Road Traffic forecasts in the light of the measures in this White Paper;
- lorry weights and the development of rail freight;
- the review of transport safety arrangements;
- progress on the take-up of green transport plans;
- the new rural bus partnership fund in England;

- how to secure best value from public subsidy for the bus industry in the longer term;
- public expenditure priorities for integrated transport in the longer term;
- research, in particular with a view to gaining a better understanding of the costs and benefits of transport and how these relate to the costs faced by users.

4.5 Our new approach to transport is not the property of any one party or interest group. The *Commission* will have an independent chair and a small permanent core of members, chosen in part to represent particular interests but principally for their expertise and impartiality. It will include representatives of Scotland, Wales and Northern Ireland, someone from the science and technology community and a transport user representative. The *Commission* will be required to consult widely with providers and regulators, central and local government, regional bodies, interest groups, trade unions, business and users. It will also draw on the expertise and resources of other organisations and individuals drawn in for work on particular topics. The arrangements for dealing with devolved matters will be set out in the Scottish Integrated Transport White Paper and the Welsh transport policy statement.

4.6 The *Commission* will make recommendations to Ministers and prepare an annual report on the implementation of the new approach, including progress towards meeting targets, the impacts of key policy initiatives and priorities for further action.

Funding transport

4.7 Transport makes a significant call on the public purse – this year, for example, planned expenditure includes some £1.6 billion on railways in Great Britain, around £3 billion on local transport in England and £1.3 billion for the English trunk and motorway network. **We will ensure that public expenditure on transport is firmly directed towards delivering the *New Deal for transport*.** In addition, through partnership with the private sector, we expect to see the level of privately-financed investment in transport increase by at least a half over the next three years.

4.8 Responses to our consultation last year sought significant improvements in all modes of transport, in rail and bus services, in conditions for pedestrians and cyclists and on our roads. As Government, we have a duty to balance calls for increased public spending against the need to maintain stable and prudent finances over the economic cycle. Our Economic and Fiscal Strategy sets out our framework for future spending which will allow real current spending to grow in line with the growth of the economy, whilst enabling us to increase capital spending to double the level of net public investment as a share of GDP. Our transport infrastructure in particular will benefit from this significant boost to public investment.

4.9 But we have to determine the balance of expenditure on public services according to our economic and social priorities. Although transport investment at national level could be funded by dedicating particular streams of taxation income, as some suggest, that would inevitably restrict our ability to use that income flexibly both for transport and for other priorities such as education and health.

4.10 Our approach is to take a strategic view. That is why we carried out a Comprehensive Spending Review across government. We have matched spending to our priorities. For transport, these are to ensure that we properly maintain and manage our existing infrastructure and that we support the delivery of integrated transport locally to reduce congestion, improve the environment and increase accessibility for everyone.

4.11 Our transport policy sets the context in which roads and railways will be planned and

CHAPTER 4 — Making It Happen

operated. To improve efficiency and to reduce the impact of transport on the environment, our focus must be on the need for better management, maintenance and use of what we have. Managing any infrastructure needs a long term perspective and a degree of certainty about approach and funding, otherwise it is difficult to plan ahead and make the best use of the resources available. Recognising this, we announced in the Economic and Fiscal Strategy a major reform of the public spending rules. We have abolished the annual spending round which encouraged short-termism and inefficiency. Firm spending limits for the next three years will give us greater certainty and stability to plan and manage our programmes sensibly.

The New Deal for transport, new ways of funding

- new sources of finance to relieve the burden on the taxpayer, for example:
    - public-private partnership for London Underground, bringing in some £7 billion of investment, and for the Channel Tunnel Rail Link with some £6 billion of investment;
    - public-private partnership for air traffic services to secure future investment needs;
    - local authority airports – relaxation of public sector borrowing controls;
    - dedicated income streams from road user charging and parking levies to fund local transport packages;
    - pilot charging schemes for motorways and trunk roads;
- advice from the *Commission for Integrated Transport* on the costs and benefits of transport.

Strategic Rail Authority

4.12 We announced in our Manifesto that **we will establish a new rail authority to provide a clear, coherent and strategic programme for the development of our railways**. The Environment, Transport and Regional Affairs Committee of the House of Commons[1] has supported our plan for a *Strategic Rail Authority*, as a practical way of addressing the problems of the restructured railway. The new authority will be a statutory body with board members appointed by Ministers. It will have a strong voice for the consumer with consumer representation on its Board. It will be subject to instructions and guidance laid down by Ministers in accordance with the new integrated transport policy.

4.13 The authority will consult the devolved administrations in Scotland and Wales about the exercise of its functions as they relate to their interests and will play an active role in the new arrangements for regional planning in England. The Scottish Executive will be able to issue instructions and guidance to the *Strategic Rail Authority* for passenger rail services which both start and end in Scotland and for Scottish sleeper services. The Scottish Parliament will have legislative competence in respect of the rail functions of the Strathclyde Passenger Transport Authority/Executive.

4.14 The *Strategic Rail Authority* **will provide a focus for strategic planning of the passenger and freight railways with appropriate powers to influence the behaviour of key industry players.** This will provide a better means of influencing the use of the significant amounts of public funds which we provide to the industry. **The Authority will:**

- promote the use of the railway within an integrated transport system;

---

1 Third Report of the Environment, Transport and Regional Affairs Committee "The Proposed Strategic Rail Authority and Railway Regulation", Stationery Office, March 1998. ISBN 0-10-221498-0.

- ensure that the railways are planned and operated as a coherent network, not merely a collection of different franchises;

- work closely with local and national organisations, including local authorities, Regional Planning Conferences, Regional Development Agencies, transport operators and the Highways Agency and the equivalent organisations in Scotland and Wales to promote better integration;

- participate actively in the development of regional and local land use planning policies, and ensure as far as possible that decisions on the provision of rail services dovetail with these policies;

- ensure that rail transport options are assessed in a way which constitutes good value for money and optimise social and environmental gains;

- take a view on the capacity of the railway, assess investment needs, and identify priorities where operators' aspirations may conflict with one another;

- promote the provision of accessible transport for disabled people;

- keep under review and advise Government on the contribution that the railway can make to sustainable development objectives;

- draw up policies and criteria for any future framework for competition between passenger train operators.

4.15 The *Strategic Rail Authority* will not be constrained by the Franchising Director's current narrow focus on the passenger railway. **It will support integrated transport initiatives and provide for the first time a clear focus for the promotion of rail freight.** The *Authority* will ensure that freight interests are given due weight both in long term planning and day-to-day decisions. It will take over from the DETR the function of administering the rail freight grant scheme in England.[2]

4.16 The authority will take over responsibility from the Office of Passenger Rail Franchising (OPRAF) for the management of passenger rail franchises and the administration of subsidy for passenger services. **The *Strategic Rail Authority* will become the main regulator of passenger network benefits** (ie the benefits of an integrated network of train services, including such things as through-ticketing and passenger information), thus avoiding the present confusion about the respective roles of the Office of the Rail Regulator and OPRAF. New sanctions will enable stronger and more timely action to be taken against operators who breach their contracts or licences. Section 55 of the Railways Act 1993 will be amended to make it less cumbersome and to remove ambiguities which have emerged in practice. In future, it will be possible to impose penalties in respect of past breaches which have ceased, and to take quicker enforcement action in a way that will still be fair to operators.

Railways – fares

4.17 One of the most obvious failures of rail privatisation has been the perceived lack of a clear, understandable national fare structure. Some key fares are regulated by the Franchising Director and from 1999 until 2003 fare increases will be restricted to Retail Price Index minus 1% – a fall in real terms. But many popular fares such as APEX, cheap day singles and returns are set entirely at the discretion of the individual train operator. Although train operators have introduced some new and innovative fares, this has led to a multiplicity of different and frequently changing fares for similar services with, in some cases, complex and varied conditions, for example in relation to advance booking.

---

2   The Scottish Executive will be responsible for administering freight facilities grant and track access grant in Scotland, and the National Assembly for Wales will administer freight facilities grant in Wales.

CHAPTER 4 — Making It Happen

4.18 The controls – and the absence of controls – are a consequence of legally binding franchise agreements inherited from the previous Government. There is little practical scope for altering them in the short term. But when opportunities arise for negotiating franchises, the new *Strategic Rail Authority*, guided by Ministers, will ensure that arrangements are made so that train operators structure and market their fares to offer value for money for their customers, and to reflect the fact that the railway is a national network which needs to be marketed accordingly and in a way which encourages people to switch from car to train.

Railways – better services, accountable to passengers

4.19 Fares are only one of many key decisions that are currently reflected in franchise agreements with train operators. A number of franchises expire in 2003/4. **In seeking new operators, the *Strategic Rail Authority* will have the opportunity to specify service levels and passenger benefits which fully reflect our integrated transport policy.** We will retain the capability for the public sector to take over franchises as a last resort, for

Franchised Passenger Train Operating Companies (GB)

Franchise Operator	Owner	End date	Length of franchise
Great Western Trains Co. Ltd	Firstgroup plc	Feb 2006	10 yrs
South West Trains Ltd	Stagecoach Holdings plc	Feb 2003	7 yrs
Great North Eastern Railway Ltd	Sea Containers	Apr 2003	7 yrs
Gatwick Express Railway Co. Ltd	National Express Group plc	May 2011	15 yrs
Midland Main Line Ltd	National Express Group plc	Apr 2006	10 yrs
Connex South Central Ltd	Connex Rail	May 2003	7 yrs
LTS Rail Ltd	Prism Rail plc	May 2011	15 yrs
Chiltern Railways Co. Ltd	M40 Trains Ltd	July 2003	7 yrs
Connex South Eastern Ltd	Connex Rail	Oct 2011	15 yrs
Cardiff Railway Co. Ltd	Prism Rail plc	Apr 2004	7 yrs 6 mths
Wales & West Railway Ltd	Prism Rail plc	Apr 2004	7 yrs 6 mths
Thames Trains Ltd	Go-Ahead Group plc	Apr 2004	7 yrs 6 mths
Island Line Ltd	Stagecoach Holdings plc	Oct 2001	5 yrs
Anglia Railways Train Services Ltd	GB Railways Group plc	Apr 2004	7 yrs 3 mths
Great Eastern Railway Ltd	Firstgroup plc	Apr 2004	7 yrs 3 mths
West Anglia Great Northern Railways Ltd	Prism Rail plc	Apr 2004	7 yrs 3 mths
CrossCountry Trains Ltd	Virgin Rail Group Ltd	Apr 2012	15 yrs 3 mths
Merseyrail Electrics Ltd	MTL Trust Holdings Ltd	Mar 2004	7 yrs 2 mths
North Western Trains	Firstgroup plc	Apr 2004	7 yrs 1 mth
Northern Spirit Ltd	MTL Trust Holdings Ltd	Apr 2004	7 yrs 1 mth
Central Trains Ltd	National Express Group plc	Apr 2004	7 yrs 1 mth
Thameslink Rail Ltd	Go-Via (Go-Ahead Group plc and VIA GTI)	Apr 2004	7 yrs 1 mth
Silverlink Train Services Ltd	National Express Group plc	Oct 2004	7 yrs 8 mths
InterCity West Coast Ltd	Virgin Rail Group Ltd	Mar 2012	15 yrs
ScotRail Railways Ltd	National Express Group plc	Apr 2004	7 yrs

example, if there are no acceptable private sector bids. The *Strategic Rail Authority* will in due course assume the British Railways Board's responsibilities.

4.20 We intend to forge a new relationship with the passenger railway, for the benefit of the people that it exists to serve. The *Strategic Rail Authority* will be our prime vehicle for this, combining pragmatism with a strategic view – the attitude which will henceforth characterise our dealings with the franchised railway. We are willing to consider renegotiation of existing franchises but only where this would secure a dividend for the passenger in terms of improved investment and services as well as value for public money. The performance of existing franchises will be a key criterion for future franchise awards. **We will expect to see in all new franchises – and in any that are renegotiated – more demanding performance standards for train operators and arrangements which enable passengers to hold operators to account for the services they run. Passengers must in future have a greater voice in train services which are paid for with their fares and their taxes.**

Railways – the passenger's voice

4.21 **We want passengers to have a real say in the railway system which we are creating.** We will transfer responsibility for the statutory passenger representative bodies (the Central Rail Users Consultative Committee and the Regional Committees – CRUCC and RUCC) from the Rail Regulator to the *Strategic Rail Authority* as recommended by the Environment, Transport and Regional Affairs Committee. We value the work of the CRUCC and RUCCs and recognise the need to make more use of the Consultative Committee network. But we also want the Committees to co-operate with bus user representative bodies such as the National Bus Users' Federation and to contribute jointly to the development of the regional transport strategies which will form part of Regional Planning Guidance (described later in this Chapter) and more generally to provide a voice for the passenger in the regions. We will consider how best the Committees can give consumers an effective and powerful voice. We will also ensure that they include a wide cross-section of passengers as recommended by the Select Committee.

The Rail Regulator: infrastructure investment

4.22 The rail industry will need an element of stability and certainty if it is to plan its activities effectively. But the *Strategic Rail Authority* should not be responsible for Government subsidy to the industry and at the same time for setting the charges which form such a large part of the subsidy bill. There will remain a number of key tasks that are best left to an independent Rail Regulator. We will enhance the Regulator's existing duties by a new duty to have regard to statutory guidance from the Secretary of State on his broad policy objectives for the passenger and freight railway.

4.23 The Regulator's functions will include setting the charges for track and station access, and for any investment required by the *Strategic Rail Authority*. He will continue to assess whether Railtrack is delivering the investment and maintenance programmes underpinning the charges, and to be responsible for securing compliance with Railtrack's network licence. The Rail Regulator will continue to have concurrent powers, including those to be granted under the Competition Bill. These arrangements will take account of general regulatory practice emerging from our Green Paper[3] on utility regulation, where this is appropriate.

---

3 "A Fair Deal for Consumers: Modernising the Framework for Utility Regulation", published by The Stationery Office, March 1998.

4.24 The Environment, Transport and Regional Affairs Committee drew attention to the importance of the Rail Regulator's forthcoming review of Railtrack's access charges, which will determine the amount which Railtrack can charge train operators from 2001 onwards. **We look to the Regulator to address in his review both the level and structure of charges. This should include not just an assessment of how much Railtrack should be paid but also of mechanisms for payment.** We must ensure that Railtrack has adequate incentives to perform effectively and efficiently, meeting the needs of customers and funders and meeting its obligations to make the best use of the existing network and where appropriate to develop that network.

The Rail Regulator: rolling stock leasing companies

4.25 The National Audit Office (NAO) in its report on the Privatisation of the Rolling Stock Leasing Companies noted that the over-riding objective of the then Government was to secure the sale of the companies as soon as possible in 1995. The NAO also noted that the chosen timing of the sale probably had an adverse impact on proceeds. The absence of effective controls over the railway rolling stock leasing companies has been a frequently expressed and long standing concern. Successive Transport Select Committees have recommended action to strengthen regulation of these companies.

4.26 In the light of these concerns, we asked the Rail Regulator in January of this year to report on the operation of the rolling stock leasing market. His report, published on 15 May 1998, rejects the previous Government's assumption that the leasing companies do not have market power. The Competition Bill, now before Parliament, would provide substantial protection against abuse of that power. But the Regulator also recommends negotiation with the leasing companies of rules of conduct, to back up competition legislation. **We agree that a concordat between the Regulator and the companies is the minimum necessary to protect the public interest and have asked him to enter discussions with them. We will review the need for further action, including regulation, in the light of the outcome.**

Investment in rail

4.27 Passenger train operators are required under the terms of their franchises to make substantial investment, notably in ordering new rolling stock. So far, these have translated into commitments for nearly £1.6 billion worth of new or re-bodied stock as well as other contractual commitments to improve services to passengers, for example, through station improvements, which will represent additional investment. Moreover, voluntary commitments for a further £230 million of rolling stock have been entered into since franchise award. The negotiation by the Franchising Director of "passenger dividends" where control of franchises has changed has led to commitments to yet more rolling stock, to station improvements and to more frequent services. We are determined to ensure that passengers and taxpayers get full value for the government subsidy provided to the railways. We want more capital investment for the benefit of passengers; more accountability to passengers and, through the *Strategic Rail Authority*, to Government; and more emphasis on network benefits and integration. As noted earlier, re-negotiation or re-letting of franchises will provide opportunities to put these principles into practice.

4.28 The Rail Regulator has identified the need for tighter controls on the implementation of Railtrack's investment programme for the 10,000 route miles of railway and associated infrastructure for which it is responsible. Last year, he secured a licence modification which strengthens his powers to enforce the implementation of Railtrack's Network Management Statement. This will be important in ensuring the rate of investment which is needed to secure the best use of this important national asset. Our proposals for stronger and more timely sanctions under Section 55 of the Railways Act 1993 will strengthen the powers of the Rail Regulator, as well as enabling the *Strategic Rail Authority* to take action against operators who breach their contracts. **Stronger enforcement powers will ensure that the modified Railtrack licence provides more effective and accountable regulation.**

4.29 We need to take a strategic, network-wide view of the development of the railway and its contribution to an integrated transport policy. In November 1997 we issued revised Objectives, Instructions and Guidance to the Franchising Director, requiring him to provide an assessment of Railtrack's investment plans, as set out in the Network Management Statement, as part of a wider review of the type and level of service that the rail network should provide. The assessment is considering:

- whether the taxpayer is getting value for money for the track and station access charges already committed to Railtrack, in terms of a better quality network;

- whether Railtrack is doing enough to facilitate the progressive improvement in passenger services and facilities, and increase in the number of rail passengers, consistent with Government policy;

- evidence of bottlenecks on the rail network and the action Railtrack is proposing to tackle them;

- the actions Railtrack proposes to take, with train operators, to improve overall levels of train operating performance.

4.30 The Franchising Director is also contributing to the Rail Regulator's own examination of whether the Network Management Statement published by Railtrack in March 1998 is compliant with Railtrack's stewardship obligations under the terms of its licence. The Regulator noted at the time of publication that the statement provided passengers and freight customers with greater detail about Railtrack's plans but that the statement, as it stood, contained very few firm commitments to deliver significant improvements across the railway network which passengers and customers could recognise as such. As a first step, he is therefore finding out from train operators and funders of the railway whether Railtrack's statement meets their reasonable needs, as required by Railtrack's licence.

4.31 To encourage further investment in the rail network, **we are providing the Franchising Director with additional funds aimed at supporting new investment proposals that produce significant wider benefits for both integration and modal shift. This will be distributed through two new schemes; the Infrastructure Investment Fund and the Rail Passenger Partnership scheme.**

4.32 The Infrastructure Investment Fund will support strategic investment projects aimed at addressing capacity constraints at key infrastructure 'pinch-points' on the existing rail network. These projects will supplement the commercial infrastructure investment undertaken by Railtrack and will help to ensure that sufficient capacity is available both for existing demand and for new demand arising from initiatives to encourage more passengers and freight onto the railway.

4.33 The Rail Passenger Partnership scheme is designed to encourage and support innovative proposals at the regional and local level that develop rail use and promote modal shift. Support will be targeted on proposals that offer the greatest opportunities for modal shift and integration with other modes, for example those that increase accessibility for disabled people and more generally improve the attractiveness of rail to both existing and potential new users. Support for these projects will help to increase further the quality of service offered by local and regional rail.

4.34 The Environment, Transport and Regional Affairs Committee recommended that the new Rail Authority should have the power to offer guarantees to existing franchisees which are proposing to invest in new rolling stock, that any subsequent franchisee would be required to take over the lease for that stock. The Franchising Director already has this power and it will be transferred to the *Strategic Rail Authority*.

4.35 The Select Committee also expressed concern about the disposal of land owned by the British Railways Board and made a number of recommendations. We have already taken steps to ensure that Railtrack, rail businesses and local authorities, are kept informed by British Railways of land sales so that they have an opportunity to acquire sites which could be used in developing the rail network. However, in view of the importance of ensuring that sites which are of potential value to the passenger or freight railway are identified, we have agreed with the British Railways Board that it should suspend land sales immediately while it conducts an audit of the remaining sites. The Board will discuss with key players in the industry its plans for the future and, in the light of that, report to the Government on any sites which have a realistic prospect of use for transport purposes in the foreseeable future. We shall then ensure that Railtrack, the rail businesses and relevant local authorities have ample opportunity to bid for those sites.

4.36 In order to secure increased use of rail freight, we have taken action to boost the take up of freight grants, which are paid to tip the balance in favour of rail haulage where the environmental benefits justify that. 1997/98 saw a substantial increase in expenditure on freight grants compared to previous years with nearly £30 million being spent, nearly double that in 1996/97. Over four million lorry trips will have been saved as a result of grants awarded since the scheme began. In order **to support our new emphasis on the role of rail freight, we have substantially increased the funds available for these grants.**

4.37 We will also look at the contribution that inter-modal freight terminals and 'piggyback' style operations (ie where lorry trailers are carried on rail wagons) could make to increasing rail freight's share of the market. Such developments would require significant commitment from the private sector in terms of investment in infrastructure and operations. We will consider applications from Railtrack and others for additional public investment on a case by case basis.

Lorry trailers carried 'piggyback' on rail wagons.

Investment in trunk roads

4.38 Our road network is largely complete. **Maintaining the trunk road network will be the first priority in future.** Details of our refocused investment strategy for trunk roads in England will be set out in the report of the Roads Review in England. There will be separate reports for Scotland and Wales.

4.39 Following years of inadequate funding, we have increased the resources available for trunk road maintenance in England. The Highways Agency is reviewing its contractual arrangements for maintenance to explore the scope for new partnership arrangements. The review will consider how far longer term contracts accompanied by additional risk transfer can ensure that maintenance will be carried out more effectively and efficiently. The Agency is also exploring the scope for private finance projects for the maintenance, finance and operation (MFO) of trunk roads. We will consider other ways of providing greater funding stability, and will consider how further incentives may be built into funding mechanisms to encourage the optimisation of whole life costs.

Aviation and airport regulation

4.40 Large airports inevitably exercise a degree of monopoly power over the market for air travel into and out of their areas. We will ensure that the system of economic regulation of airports continues to promote the interests of airport users, both airlines and air passengers. In preparing a new airports policy White Paper (see Chapter 3), we will consider how airport regulation should support our wider transport policy objectives.

4.41 Our Green Paper on utility regulation includes proposals to maintain and enhance the effectiveness of the system by bringing airport regulation into line with the model which applies to other utilities and by granting the CAA, as airports regulator, concurrent powers with the Director General of Fair Trading under the Competition Bill. We have also sought views on the extent to which other reforms recommended for the energy, water and telecommunications utilities might be applied to airports, such as giving the CAA a primary duty to protect the interests of consumers and the creation of a duty on regulators to have specific regard to Ministerial guidance on environmental objectives. In the light of responses to the consultation, we will bring forward proposals for legislation.

4.42 We will continue to promote the interests of our successful UK aviation industry and both passengers and freight users through the negotiation of international air services agreements. Our aim is to achieve further liberalisation of international air services wherever possible in our bilateral aviation negotiations with other countries. We will only support the European Commission taking over responsibility for aviation negotiations when it can demonstrate that it could do better than Member States acting on their own.

4.43 We wish to see the liberalisation of transatlantic services, our largest aviation market outside the EU. However, such liberalisation must be on the basis of fair competition, including effective access to the large US internal market, and with adequate protection for small and new entrant carriers against abuse of market power.

4.44 The proposed alliance between British Airways and American Airlines would provide effective access to the US internal market for the UK's largest carrier, but has yet to be authorised by the competition authorities. Authorisation is at least in part dependent on the liberalisation of the UK-US market.

Investment in aviation

4.45 We believe that the level of investment and efficiency that we need in our National Air Traffic Services (NATS) can best be achieved through a partnership between the public and private sectors. We therefore recently announced proposals for a partnership of this kind to help NATS finance the investment it needs to operate effectively and to mobilise private sector resources. Our preference is that 49% of the shares, and a golden share, are held by the Government; and 51% by private investors, including employees. We will consult further on the implementation of this decision, which will require legislation.

4.46 Our aim is to guarantee the highest safety standards. The CAA is currently responsible for the provision of air traffic control services though NATS and for regulating the safety of those services. We will bring forward proposals to ensure that air safety regulation is conducted separately from NATS and for economic regulation of air traffic control services. We will ensure that safety

CHAPTER 4   Making It Happen

regulation is independent, open and transparent and that the industry and its employees can play a full part.

Trust Ports

4.47 The diversity of ports in the UK and the competition between them offers benefits to their customers and to the wider economy. We believe there is a role for the private sector and for trust ports – established under local Acts of Parliament and run on a non profit-making basis for the benefit of all port users and wider local/regional interests. **We withdrew the previous Government's plans to force the privatisation of trust ports**.

4.48 We are currently reviewing the role and status of trust ports in Britain (of which there are over 90), in particular in relation to their operations, economic activities, accountability and the constraints of their statutory powers and duties. We will consult with the industry and other interested parties and announce our conclusions in due course.

Devolution

4.49 Different parts of the UK have differing transport needs. Scotland, Wales and Northern Ireland will be able to consider their own transport priorities under the new arrangements for a Scottish Parliament, a National Assembly for Wales and an Assembly for Northern Ireland. Our plans to devolve greater powers on transport and other matters will improve local accountability and democracy, helping to ensure that solutions reflect local needs and circumstances.

Regional action

Integrating transport and planning in the English regions

4.50 Our proposals for modernising the planning system in England highlight the importance of planning at the regional level. Regional planning conferences or similar groups of local planning authorities will have a key role in advising on the most sustainable way of meeting the demand for new housing in their regions. Earlier this year, we consulted on a package of proposals for reforming Regional Planning Guidance (RPG). **A key proposal is for RPG to include a regional transport strategy.** Our proposals for improving RPG and promoting greater ownership have been widely welcomed. In the light of the responses to this consultation, we will publish draft guidance (PPG11) setting out the new arrangements in more detail, including the scope of the regional transport strategies and how they will be prepared.

Easier connections: bus and tram interchange in Manchester

Regional transport strategies

4.51 In England, regional planning conferences or similar groups of local authorities working with the Government Offices for the Regions, in partnership with RDAs, will have direct responsibility for preparing the new RPG in draft and for consulting widely on it. This replaces the current arrangement where the planning conferences simply give advice to the Secretary of State. It means that regional conferences will be responsible for the development of long term regional transport strategies, giving people a greater say in what happens in their region.

4.52 The conference's proposals will be discussed at a public examination – independently chaired – before going to the Secretary of State for final approval. To work effectively, these transport strategies will need to reflect our integrated transport policy. They will also need to be drawn up in close consultation with the relevant Regional Chamber, especially if it is a designated chamber[4], and with representatives of passengers and other transport users. In approving RPG, the Secretary of State will need to be satisfied that the transport strategy does not conflict, without good local reason, with national policies.

4.53 The regional conferences will use RPG to integrate the planning of major new development at the regional level and the identification of regional transport investment and management priorities. In doing so, the conferences will need to consider including in RPG:

- public transport accessibility criteria for regionally or sub-regionally significant levels or types of development, to be set out in development plan policies to guide the location of development;

- guidance for development plans on the approach to be taken to standards for off-street car parking provision, relating these to accessibility by public transport;

- a strategic steer on the role of airports and ports in the region in the light of national policy;

- regional priorities for transport investment and management to support the regional strategy, including the role of trunk and local roads;

- traffic management issues which require consideration either regionally or sub-regionally;

- guidance to local authorities on the strategic context for introducing measures such as road user charging and parking levies.

4.54 In developing the regional strategy, conferences will also have to liaise closely with transport operators and infrastructure providers in their regions, the Highways Agency and the *Strategic Rail Authority*.

Regional Planning Guidance and Trunk Road Planning

- to develop an effective integrated transport policy at the regional level, decisions about trunk road planning should be set in the context of the transport network as a whole;

- the definition of long term regional priorities for transport improvement and management in Regional Planning Guidance must flow from an appraisal of the realistic options available and from an understanding of the role of transport in sustainable regional development;

- we will look to conferences of local authorities to work with their regional partners to consider the objectives and, in broad terms, the priorities for managing and improving trunk roads which are key to delivering the regional strategy;

- our investment strategy for trunk roads will be consistent with the priorities set out in Regional Planning Guidance.

---

4   designated by the Secretary of State under the powers proposed in the Regional Development Agencies Bill.

CHAPTER 4 Making It Happen

4.55 Government Offices for the Region and the Highways Agency will contribute to the work of the planning conferences, including on trunk road issues, ensuring that the studies that emerge from the English Roads Review dovetail with the needs of the regional conferences.

4.56 Regional conferences will be expected to take account of our integrated transport policy, including national guidance on the role and nature of trunk roads, including the core network, and published proposals for improvement. The report of the Roads Review, the new PPG11 and a revision of PPG13 will provide guidance on these matters.

4.57 RPG will also need to take into account the implications at a regional level of EU policies, including the evolving European Spatial Development Perspective (ESDP). We are keen to see the completion of the ESDP and the promotion of the good practice it contains for considering the combined impact of policies, such as transport and the environment, on an area or region which transcends Member States' borders. The final document, reflecting consultation, will be produced in 1999.

Thames Gateway regeneration:

- to support regeneration in the Thames Gateway area, we agreed that the A13 Design, Build Finance and Operate (DBFO) project should go ahead;

- the terms of the DBFO invitation to tender includes incentives to provide a good road service for buses and commercial vehicles without encouraging car commuters;

- an integrated package of new river crossings, including better public transport links and local road crossings, could also help to stimulate regeneration and ease congestion. We are considering the options and the way forward so that decisions can be taken by the Greater London Authority.

Role of Regional Development Agencies

4.58 Our White Paper "Building Partnerships for Prosperity"[5] said that the new Regional Development Agencies (RDAs) would have a role in influencing the development of integrated transport strategies. We see that role being secured through them being key partners in the preparation of Regional Planning Guidance.

4.59 The creation of RDAs will mark a step change in efforts to improve the economic and social well being of the regions. They will have powers to foster economic development and regeneration, promote business efficiency, investment and employment and contribute to sustainable development. They will inherit the regional regeneration programmes of English Partnerships and the Rural Development Commission and will take on the administration of the Single Regeneration Budget Challenge Fund. They will also be responsible for decisions on grants for site access roads to aid economic development.

4.60 Regional transport strategies will need to take account of RDA strategies for sustainable economic development and regeneration. Both strategies will be developed in close collaboration so that the transport implications of the economic strategy can be reflected in Regional Planning Guidance and vice versa. Regional Development Agencies will have an important role in identifying development opportunities and promoting the necessary infrastructure to support them. They will also want to take account of the contribution which ports and airports make in their regions.

---

5 "Building Partnerships for Prosperity – sustainable growth, competitiveness and employment in the English regions", December 1997, Cm 3814.

Regional action

Integrated transport in London

4.61 There is currently no single body in overall charge of co-ordinating transport in London. There are many different players – central Government, boroughs, nationalised industries, quangos, private sector operators – and a variety of ad hoc arrangements, but no one can pull all their initiatives together. In London this fragmentation is a serious obstacle to pursuing the integrated approach which we want to see.

Giving buses priority on a London Red Route.

4.62 That is why we propose to give **a major transport role to the new Greater London Authority (GLA), headed by a directly elected Mayor**. The Mayor will produce an integrated transport strategy for London, covering nearly seven million people who live there as well as the millions more who travel to work or visit the capital. This will be a wide-ranging strategy, covering all forms of transport to, from and within London. With responsibilities for strategic land use planning and economic development, the Mayor will be able to ensure that transport policy is integrated with these other important policies.

New arrangements for integration in London

- *integrating transport:* the Mayor will produce a transport strategy covering all modes of travel to, from and within London. Responsibility for underground, bus and strategic roads will be brought together;

- *integration between transport and the environment:* the Mayor will have a statutory duty to promote sustainable development, and specific environmental functions including the production of an air quality strategy for London, a duty to produce reports on London's contribution to national climate change targets and powers and duties in relation to noise;

- *integration of transport and land use planning:* the Mayor will produce the spatial development strategy for London, covering all strategic land use planning issues, transport policy and provision, economic development and regeneration, housing, retail development, town centres, and protection and enhancement of the environment;

- *integration with other policies:* the London Development Agency will be an arm of the GLA, enabling the Mayor to integrate policies for economic development with transport and planning polices. With an overview of both transport and the Police Authority's strategy, the Mayor will be uniquely placed to develop a vision of the traffic policing needs of London.

4.63 We have defined the responsibilities of the GLA in our White Paper "A Mayor and Assembly for London"[6] so that it can focus on issues that need to be tackled on a London-wide basis. It will not assume responsibilities which can be discharged at the local level.

---

6   "A Mayor and Assembly for London – The Government's Proposals for Modernising the Governance of London", March 1998, Cm 3897.

CHAPTER 4   Making It Happen

4.64 The transport strategy will be implemented through a new executive body, Transport for London (TfL), directly accountable to the Mayor. TfL will run or manage transport services in London on a day-to-day basis.

4.65 The Mayor's transport strategy will cover London-wide strategies for the bus and cycle priority network, and for freight, parking and walking. The London boroughs will frame local implementation plans to give effect to these strategies and the GLA will have powers to ensure that borough plans are in accordance with its overall strategy for London.

4.66 Our plans for powers to introduce road user charging and a levy on parking spaces (see later in this Chapter) will give the Mayor important tools for tackling congestion and air pollution, especially in central London. This would also generate extra revenue which could be used for the improvements in public transport that would be essential to cater for a significant modal shift in London. With these powers, and the direct responsibilities listed above, we believe that the Mayor will be able to formulate and deliver a truly integrated transport policy for London.

4.67 We are committed to devolving decisions to the most appropriate local level. We do not expect Government to interfere unless the Mayor's transport strategy is inconsistent with a published statement of our national transport policy, in a way that has an adverse impact beyond London. In these circumstances, the Government will have a reserve power to direct the Mayor to amend the strategy.

Investment in London Underground

4.68 **Our radical and innovative public-private partnership for London Underground is intended to bring about some £7 billion of investment in the system over 15 years, whilst retaining a publicly owned and publicly accountable network.** London Underground will continue to operate the network and will invite bids from contractors to modernise and maintain the infrastructure and trains. The elements of the existing system that passengers value, such as Travelcard, integrated ticketing and the high priority given to safety, will remain the responsibility of London Underground, whilst the main cause of complaints – long-standing underinvestment – will be addressed.

Investing in London Underground: a new Northern Line train.

4.69 To meet the immediate investment needs of the Underground system, we have given London Transport an additional £365 million over the next two years for core Underground investment and for preparing the new public-private partnership. This will bring total investment in the current network to £1 billion over the next two years, allowing many important projects to go ahead.

Regional action

Docklands Light Railway – the integrated railway

Bank - main interchange with London Underground

Stratford - multi-modal interchange with the proposed international rail station, London Underground, National Railways and buses

Canning Town - multi-modal interchange with London Underground, National Railways and buses

London City Airport - international rail-air interchange

Cutty Sark - multi-modal interchange with bus link to Millennium Dome and river services

Lewisham - integrated ticketing facilities with National Railways and buses

This map shows the route of Docklands Light Railway (DLR) and key links with the railway. Stations highlighted show important examples of how DLR integrates with other modes of public transport.

Key
- Docklands Light Railway
- DLR routes planned or under construction
- Proposed cable car
- Millennium Transit
- Jubilee line extension
- East London line
- North London line
- Railtrack lines
- River boat piers

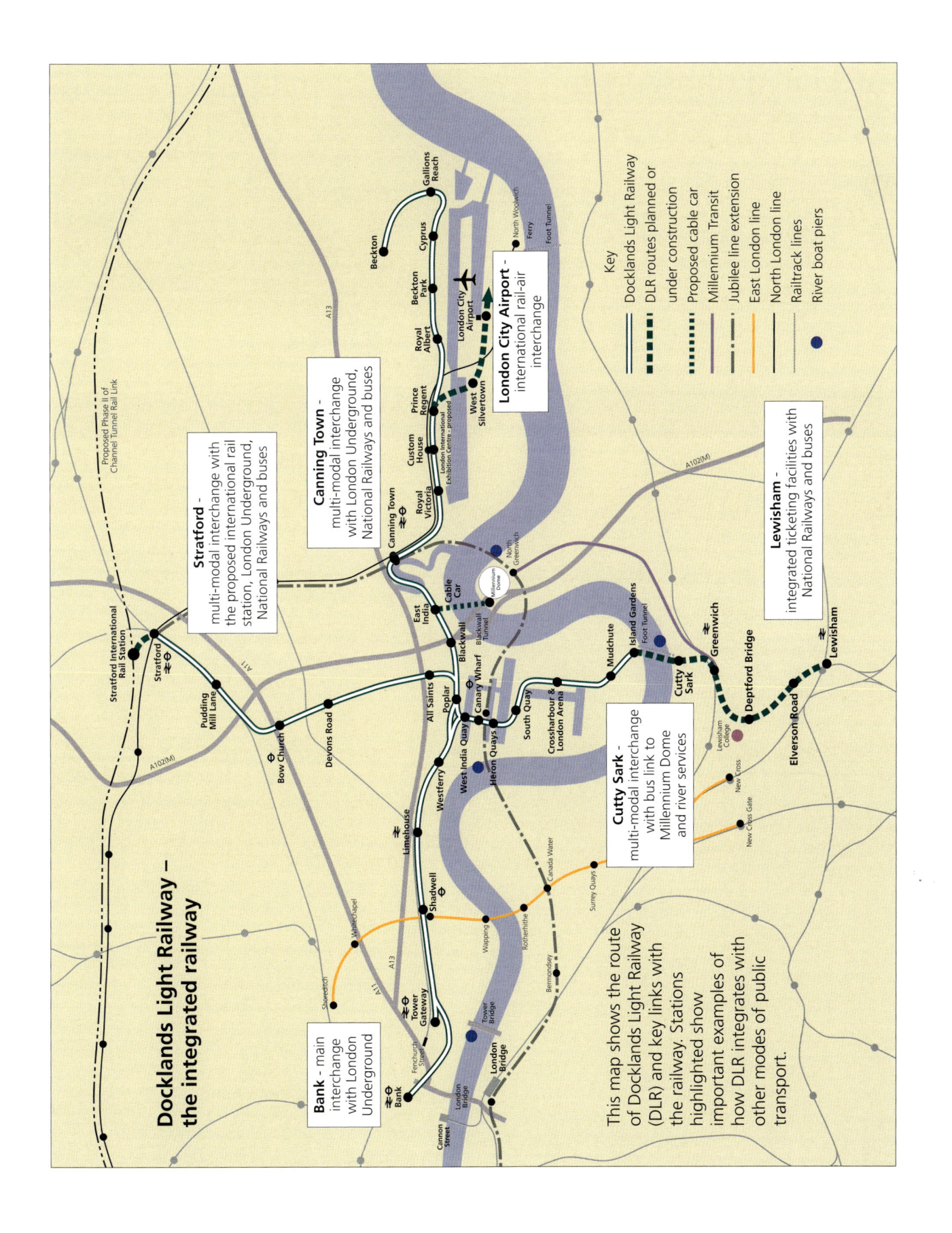

107

CHAPTER 4   Making It Happen

Integrated transport in London – in practice

4.70 **We want to use the next two years to build a strong foundation for the Mayor and GLA to tackle London's problems.** London provides an excellent illustration of what integrated transport can mean in practice.

River boat services on the Thames.

Integrated transport in London – in practice

Improvements for pedestrians

- a key objective for the Traffic Director for London is to improve conditions for pedestrians. More than 300 new or improved crossings have been provided so far as part of the Red Route programme;

World Squares for All

- aims to achieve better access and enjoyment of the ceremonial heart of London for pedestrians, visitors and tourists; and a better setting for the historic buildings and public spaces around Trafalgar Square, Whitehall, Parliament Square, balanced with the need to maintain effective bus operations and to avoid creating unacceptable levels of traffic congestion. A masterplan for the area has been produced. It includes proposals for:

    - pedestrianisation of parts of Trafalgar Square and Parliament Square;

    - a range of imaginative measures to enhance the setting and character of the whole area;

    - improving pedestrian access and safety in the Squares and on the surrounding streets.

The London Cycle Network

- local authorities are developing a comprehensive cycle network across London providing safe, convenient routes incorporating cycle lanes, protected crossings and shared paths in open spaces and parks;

- this also provides safer routes to schools and helps cyclists at major road junctions.

The London bus priority network (LBPN)

- LBPN will comprise 540 miles of the most heavily used bus routes, being developed by the local authorities and London Transport;

- involves some 1,200 new schemes such as bus lanes, bus stop clearways, measures to ensure that low-floor buses reach the kerb at bus stops and bus priority at traffic lights;

- focus will be on completing entire bus routes which deliver benefits to passengers at the earliest date. Six whole routes have been identified for comprehensive improvement this year.

Improving transport interchange

- major schemes underway include Croydon Tramlink's connections with East and West Croydon stations; Stratford Regional Station,

Regional action

Improvements for pedestrians: crossing at Baker Street

Integrated transport in London – in practice

linking the Jubilee Line to the Central Line, Docklands Light Railway and overground rail lines and the new bus station;

- improvements are being made to bus access at Willesden Junction, Tottenham Hale and Seven Sisters stations and to bus/tube interchange at Wood Green station;

- BAA has invested some £450 million in the Heathrow Express. The service will cut journey times from London to Heathrow airport to 15 minutes and is expected to carry some six million passengers in year one and to account for around 15% of airport passengers by 2001;

- Docklands Light Railway integrates with other modes of transport (see map), the proposed extension to London City airport would, its promoters claim, offer a direct rail service with the potential to take a significant number of journeys off the road at a cost of some £35 million.

Travel information in London

- London Transport's Travel Information Service provides;

  - 24-hour Call Centre telephone advice on timetables and itineraries for all modes;

  - Travel Information Centres at major locations such as mainline stations and elsewhere;

  - interactive terminals giving direct access to a computerised journey planning system at a number of trial locations in London, including shopping centres, bus stations, and tourist offices;

- Countdown system provides 'real-time' information signs at the busiest bus stops to inform passengers when the next bus will arrive. It is planned to extend the system throughout London.

Traffic management and parking guidance

This guidance to London local authorities, the Traffic Director for London and the Highways Agency, which we published earlier this year, encourages:

- a more strategic approach to parking in London, with a more determined use of parking charges and controls, such as Controlled Parking Zones;

- greater emphasis on measures to assist buses, cyclists and pedestrians, including new aims for the Red Route Network;

- recognition of the needs of all road users, especially people with disabilities or difficulty with walking;

- better interchange between modes, especially from bus and car to rail and underground, and from public transport to walking.

CHAPTER 4   Making It Happen

Role of Passenger Transport Authorities

4.71 The six English Passenger Transport Authorities and their Executives (PTAs/PTEs) (ie Greater Manchester, Merseyside, South Yorkshire, Tyne and Wear, West Midlands and West Yorkshire) and Strathclyde in Scotland are responsible for securing public transport services for some 14 million people in major urban areas outside London. They are well placed to **play a leading role in delivering integrated transport objectives in places which face some of the most serious environmental and congestion problems outside London.** In doing this, they will need to build on and extend existing joint working arrangements and partnerships with highway authorities, transport operators and other organisations in their areas. In particular, the English PTAs will need to work closely with the district councils in their areas to produce joint *local transport plans* so that the highway authorities' plans support the PTA strategy.

PTA/Es are important in developing integrated transport in metropolitan areas through:

- providing a more strategic approach to passenger transport issues in urban areas where there is a heavy reliance on public transport;
- securing tendered bus services;
- as joint signatories to rail franchise agreement, specifying and funding local rail services;
- a strong role in promoting integrated public transport services;
- close joint working with highway authorities and others, for example, in working up package bids with district councils which provide the framework for decisions on bus lanes and other priority measures.

4.72 We believe that many of the improvements we want to see can be achieved within existing powers. In order to demonstrate what can be achieved through voluntary co-operation, **Greater Manchester PTA, transport operators and the district councils in its area have developed a pilot project.** We will be looking to PTAs elsewhere to develop similar projects. We will monitor these projects closely so that the experience can inform future initiatives.

Greater Manchester – integrated transport pilot

- Greater Manchester Passenger Transport Authority has been working with public transport operators to deliver improvements across all modes – bus, rail and Metrolink;

- information and advice is available through a network of 'Travelshops', a telephone information bureau, printed timetables, network maps and departure time displays at bus stops. Travelshops also sell bus and rail tickets;

- A wider range of initiatives is now in hand to:

    - provide comprehensive publicity and information at stops and stations, and on the internet;

    - offer a full range of modal and multi-modal ticketing;

    - agree common dates for bus service changes;

    - increase substantially the number of bus shelters;

    - identify strategic routes for high frequency bus services, assisted by bus priority measures.

Local action

Local transport plans

4.73 **New *local transport plans* will be a centrepiece of our proposals.** Local authorities outside London will set out in these plans their proposals for delivering integrated transport over a five year period. The detailed arrangements in Scotland, Wales and Northern Ireland will be set out in the Scottish Integrated Transport White Paper and the transport policy statements for Wales and Northern Ireland.

Local transport plans in England

- *local transport plans* will be key to the delivery of integrated transport locally;
- local authorities will draw up 5 year plans, consulting widely with local people, businesses, transport operators and community groups;
- will include future investment plans and propose packages of measures to meet local transport needs.

The plans will:

- cover all forms of transport;
- co-ordinate and improve local transport;
- set out strategies for promoting more walking and cycling;
- promote green transport plans for journeys to work, school and other places;
- include measures to reduce social exclusion and address the needs of different groups in society;
- set out proposals for implementation, including bus *Quality Partnerships*, traffic management and traffic calming, proposals for road user charging and PNR parking charges and freight *Quality Partnerships*.

Local transport plan targets could include

- *air pollution* – to improve local air quality;
- *traffic reduction* – from the Road Traffic Reduction Act 1997;
- *cycling* – eg to increase the number of cycle trips or to increase the proportion of journeys made by cycle;
- *walking* – eg to reverse decline in walking or to increase walking journeys to school;
- *use of public transport* – eg to reverse the decline in patronage and to achieve a shift from car to bus;
- *road safety* – eg to reduce number of road casualties;
- *green transport plans* – eg for the preparation of plans by major local employers or for reducing journeys to school by car.

4.74 The plans will provide the basis for an integrated approach, closely linked with Local Agenda 21 strategies and will implement the transport aspects of development plan strategies. Regional Planning Guidance will set the regional framework for local authorities' transport plans. We will look to local authorities to build on present liaison arrangements with their neighbouring authorities (both urban and rural) and at different tiers, in the development of *local transport plans*, co-ordinating their highway authority and public transport responsibilities. Authorities will need to agree a common or complementary approach on cross-boundary issues.

4.75 In both rural and urban areas, the plan will take account of the transport and accessibility needs of local communities and business, in a way that is consistent with the new approach. Local authorities will need actively to involve local people, businesses, transport operators and other organisations such as those providing health care,

CHAPTER 4 — Making It Happen

in drawing up these plans. Guidance on the new arrangements, to be developed in consultation with local authorities and other interested parties, will reflect the importance of such local participation.

4.76 We recognise that there would be advantages in making *local transport plans* statutory and will legislate in due course. However, we are keen to introduce the new arrangements as soon as possible and will aim to do so in England on a non-statutory basis, during 1999 with the first plans covering the financial years 2000/1 – 2004/5.

4.77 Local authorities will be expected to set out in *local transport plans* their proposals for both capital and revenue expenditure on transport. To reduce central government involvement in local authority decision-making in England, we will use the new plans as a basis for an annual block allocation of credit approvals to spend on transport capital. We will expect local authorities to give due priority to cost-effective maintenance and development of their transport infrastructure to support integrated transport objectives. Consistency with the local development plan and Regional Planning Guidance will also be a factor in decisions on supporting *local transport plans*. But central government will no longer dictate specifically how resources are deployed. Instead, authorities' plans will be subject to an annual progress check. The importance of *local transport plans* as part of our strategy is reflected in the provision made for funding local transport over the next few years.

Funding bus services

4.78 **Effective local bus services will be an essential part of the new policy.** Better bus services in urban and rural areas will help to improve alternatives to the car and reduce social exclusion.

4.79 The bus industry will benefit significantly from our proposals to strengthen the role of the bus (see Chapter 3). At present around one-quarter of the seats on a bus are occupied on average. An average increase of only two passengers per bus – typically achieved by a *Quality Partnership* – could generate up to £400 million in revenue for the industry. Such initiatives can also reduce operating costs by improving reliability. In addition, the industry receives a significant level of support – around £1 billion in total this year – through fuel duty rebate (£270 million), direct subsidy (outside London, local authorities spend some £230 million on bus services), and concessionary fares (around £440 million). **Taken together, this should produce much greater financial certainty than the industry has had for many years which, together with increasing patronage, will transform the economics of the bus industry.**

4.80 The Audit Commission is currently looking at local authority revenue support in England and Wales for local transport and travel, including expenditure on locally subsidised bus services and (in the Passenger Transport Authority areas) rail services, home-to-school transport and concessionary fares. In the light of its findings and the wider development of the industry, we will ask the *Commission for Integrated Transport* to advise us on how to secure best value in the longer term from the public subsidy invested in the bus industry in support of our wider aims.

Reducing social exclusion

4.81 **We will introduce a national minimum standard for local authority concessionary fares schemes for elderly people with a maximum £5 a year charge for a pass entitling the holder to travel at half fare on buses.** This will enable elderly people, especially those on low incomes, to continue to use public transport and to use it more

---

8 "Bus Services for Rural Communities: an audit of villages in England", TAS Partnership Limited, October 1997.

often, improving their access to a range of basic necessities such as health care and shops and reducing social isolation. Local authorities will still be able to offer more generous schemes, if they wish to do so. The change will require legislation.

4.82 In urban areas, local authorities will need to explore with operators the scope for extending bus networks so that they provide better access to opportunities for work, and to goods and services, especially for those who live on remote or rundown council estates. Some bus operators have found that it makes commercial sense to offer cheaper fares in such areas. This will help to complement the action we are taking in our New Deal for Communities initiative and the Welfare to Work Scheme, under which some operators have offered discounted fares to help young, unemployed people (see Chapter 5). Local authorities may wish to consider the scope for negotiating with local transport operators, in the context of bus *Quality Partnerships*, further voluntary concessions for the less well off, especially young, unemployed people as a further means of reducing social exclusion.

4.83 Many rural areas are poorly served by public transport: some 20% of rural settlements in England[8] are estimated to have a bus service below "subsistence" levels – fewer than four return journeys a day, and no evening/weekend service. Budgetary pressures have constrained some local authorities from buying in additional services to maintain or enhance bus networks and evening/weekend services. In some cases, support is being withdrawn from socially necessary services, particularly in rural areas.

4.84 **In the March Budget, the Chancellor therefore announced a new Rural Bus Partnership fund of £45 million a year nationally to support bus services in rural areas and a further £5 million a year for our new *Rural Transport Partnership* scheme** (see Chapter 5). The arrangements in Scotland and Wales are being considered separately; in rural Scotland the support for bus services may also extend to other transport modes.

4.85 We have recently announced the arrangements for allocating £32.5 million to rural bus services in England, targeting the money on the most rural areas to provide new and additional bus services. The remaining £5 million will be allocated later this year as part of a 'bus challenge' to promote innovative local authority schemes in England, for example, to improve passenger information and services. We will monitor the effectiveness of these new measures with the help of consultants and the Traffic Commissioners. We will ask the *Commission for Integrated Transport* to advise us on future funding priorities in the light of this monitoring.

4.86 We will consult on plans for targeting the enhanced level of Fuel Duty Rebate to support rural bus services and more environmentally-friendly vehicles shortly. Taken together, **these additional funds will mark a step change in support for public transport services in rural areas.**

4.87 The need for better bus services in urban and rural areas highlights the importance of local authorities adopting clear, objective criteria for spending on public transport – to ensure best value for money. For example, in our guidance on the allocation of Rural Bus Partnership funds in England, we highlight the importance of developing public transport networks and ensuring adequate frequency, including the setting of minimum service thresholds.

4.88 We will continue to look at other ways to maintain accessibility to services and thus reduce the need to travel long distances. Planning has a role here, for example in promoting the growth of key villages or the regeneration of inner city areas, as part of a package of measures to reduce social exclusion. The local post office/village shop is very important in providing local goods and services in rural areas and we have extended the rate relief scheme to reflect this.

4.89 The village school also plays a vital role in rural communities. We want to maintain access for children to local schools: we have therefore announced that all proposals to close rural schools will be called in for decision by the Secretary of State for Education and Employment. Under both the present arrangements and the new approach we are proposing for determining school organisation proposals, there will be a presumption against closure. Information and communication technology will open up new possibilities to enrich children's learning and increase the viability of isolated rural schools. Our target is for all schools to be connected to the National Grid for Learning by 2002. By supporting local schools, these initiatives will help to reduce the need to travel in the countryside.

Funding major local transport schemes

4.90 Specific funding will have to be identified for major public transport and road schemes based largely on the present arrangements. In bidding for major schemes, local authorities in England will need to demonstrate that the scheme is necessary for achieving the objectives of the *local transport plan*, and that this cannot be done in other ways. We will use the principles behind the new approach to appraisal (described later in this Chapter) to assess bids for major local authority transport schemes. When they submit bids, we expect local authorities to demonstrate that they have explored the scope for alternative solutions that do not involve major new construction and have taken account of our strong presumption on avoiding sensitive environmental sites. We will expect local authorities to pursue public-private partnerships to finance major schemes where appropriate.

Funding local rail services

4.91 All local authorities are able to contract on a voluntary basis with train operators to provide additional services or facilities in their areas and have to fund any net cost increase arising from such services. The *Strategic Rail Authority* will develop closer relations with local authorities, offering advice on new investment schemes and working in partnership with them to promote the most attractive schemes which encourage the use of rail. In the meantime, in advising on and promoting new schemes, the Franchising Director will build on his existing criteria for assessing the cost and benefits of rail schemes. These criteria give due weight to the social and environmental benefits which railway investment can provide as part of an integrated transport policy.

Changing travel habits

Tackling congestion and pollution on local roads

4.92 Many of our towns and cities face significant levels of congestion and pollution which place a burden on business and result in a poor quality of life for people who live and work there. Some rural areas suffer from significant traffic congestion in peak holiday periods and traffic nuisance is a growing problem in the countryside more generally.

4.93 A variety of traffic management techniques can be deployed to reduce road traffic in these circumstances. As we have seen in Chapter 3, there is still scope for further and more imaginative use of such measures, combined with improvements in public transport (for example, through *Quality Partnerships*) to reduce road traffic and the associated pollution, enhancing the attractiveness of urban and rural areas.

4.94 But experience has shown that improving public transport and related traffic management measures whilst necessary are not sufficient in many cases. **We will therefore introduce legislation to allow local authorities to charge road users so as to reduce congestion, as part of a package of measures in a *local transport plan* that would include improving public transport.** The use of revenues to benefit transport serving the area where charges apply, which in many cases will mean supporting projects in more than one local authority area, will be critical to the success of such schemes.

4.95 Carefully designed schemes should reduce traffic mileage and emissions, bringing significant improvements in air quality, reducing noise and greenhouse gas emissions and relieving congestion. This will benefit pedestrians, cyclists and public transport, including more reliable and quicker bus services and more reliable delivery times for freight. Less congestion also means shorter and more reliable journey times for those who continue to drive. Charging will provide a guaranteed income stream to improve transport and support the renaissance of our towns and cities. The availability of a revenue stream will also open up the scope for greater involvement of the private sector working in partnership with local authorities.

4.96 In rural areas, road user charging is most likely to be used where there are significant problems caused by very high levels of seasonal traffic, for example, in tourist areas such as the National Parks. We would welcome proposals for such initiatives to provide the basis for pilot schemes in rural areas.

4.97 Primary legislation will be needed. Subject to that being in place, we will then work with local authorities and other interested organisations on a number of **pilot schemes individually approved by the Secretary of State (in Scotland, by the Scottish Executive). The effects of these schemes will be monitored and used to inform the design of future schemes.**

4.98 We will issue a consultation document with proposals for how road user charging schemes should operate. This will deal with different ways of implementing charges: electronic schemes, schemes where drivers must buy and display a permit and schemes using tollbooths. It will seek views on how best to ensure the active involvement of local people, business and others in the development of schemes so that proposals attract public support. We will also be seeking views on how such policies will impact on the mobility of disabled people.

Leicester Environmental Road Tolling Scheme

DETR has funded a practical trial of how drivers respond to charging, in Leicester. Key features include:

- a new, 300 space, Park and Ride site on the A47 radial corridor into the city. Comprehensive package of bus priority measures along the route from the Park and Ride site to the city centre. Peak hour journey time is now appreciably quicker by bus than by car;

- volunteers have road user charging equipment installed in their cars, and are given a travel cost account. They can use this to pay the road user charge (deducted automatically when they pass a roadside beacon) or the bus fare (regular service or park and ride);

- the project is looking at reactions to different levels of charge, and different charging periods, to see how people balance time, cost and convenience when deciding how to travel. It will also see whether the response is affected by information about air quality;

- the final report on the trial is due to be published in September 1998.

CHAPTER 4 — Making It Happen

> **London Congestion Charging Study**
>
> - a study for the Government Office for London (July 1995), investigated a range of charging levels and structures for congestion charging in Central London;
>
> - it was estimated that vehicle miles would fall by 15% and CO_2 emissions by 14.5%. Journey reliability for the remaining vehicles, notably bus operations, would improve by some 20%, and journey times would fall by a similar amount;
>
> - major improvements to public transport services and infrastructure in combination with congestion charging could increase the reduction in vehicle miles still further – to over 20%.

4.99 Following the London research, the previous government, in the 1996 Green Paper "Transport: The Way Forward" stated that it would discuss with local authorities and other interested parties how best to take matters forward, with a presumption in favour of introducing legislation to enable congestion charging to be implemented.

Charging users on motorways and trunk roads

4.100 Our proposals for legislation to allow road user charging will enable pilot schemes to be developed in a variety of circumstances. Schemes may be developed, for example, to help to meet transport and environmental objectives in urban or rural areas, or on bottlenecks on specific roads or at certain times of the day or year. **Such schemes may also be developed on trunk roads and motorways, either on a self-standing basis or as joint schemes with local authorities. Pilot charging schemes** will be individually developed and designed to take into account the local transport network, ensuring in particular that unacceptable diversion does not take place onto local roads. We will also consider for each scheme how best revenues generated may be used to provide related benefits locally which might otherwise be unaffordable, including better means of securing the environmental acceptability of transport infrastructure.

4.101 In designing further projects we will consider what lessons can be drawn from projects overseas and from those few instances where tolls are currently levied from road users in this country as a means of funding the infrastructure. Tolls have, for example, been levied for many years on estuarial crossings.

4.102 The existing powers to toll road users at Dartford are time limited. The cessation of tolling could, however, have the effect of increasing demand on the Eastern sector of the M25. We will consult on the continuation of road user charging on the Dartford crossings and how best it could contribute to delivering integrated transport policy objectives related to the M25, taking account of the need for effective traffic management and the impact on the local transport network.

4.103 On most of the motorway and trunk road network, charging schemes will in general be feasible only with full electronic technology. Further studies are required on the electronic units and on administrative support systems before they may be introduced with confidence. In particular, we need to be satisfied that such systems can cope with high volumes of traffic, travelling at motorway speeds in a way which does not produce unacceptably high error rates in charging users.

4.104 We will continue technical trials of electronic systems and carry out further research on their possible effects and how they may best be implemented. These trials will examine such issues as personal privacy, impact on different parts of society and diversion onto untolled roads. An early priority will be work to ensure that, as charging projects are introduced in different parts of the country, vehicles do not require more than one set of in-vehicle equipment. We will continue to work with the European Commission and EU Member States to ensure that the design of charging systems in Europe is compatible.

Workplace parking

4.105 Employees driving to work and enjoying free parking at the workplace account for a significant proportion of peak hour congestion. Controlling the price and availability of parking has been shown by research to be capable of reducing traffic in an area[9]. Local authorities determine the price and availability of public parking, on and off the highway. But they have little control over existing parking spaces at private business premises. They can use their development control powers to limit the amount of parking associated with new development but, in the past, development was allowed with extensive parking provision, considerably in excess of the standards advocated in current Government guidance.

4.106 We believe that new measures are needed to tackle excessive workplace parking provision at existing developments so local authorities can develop comprehensive parking management policies that support their transport and development plans.

4.107 **We will introduce legislation to enable local authorities to levy a new parking charge on workplace parking. This charge would not apply to residential parking, ie parking at or outside the home.** We propose that owners or occupiers of business premises would apply for a licence to allow a certain number of vehicles to be parked on site. The aim is to reduce the amount of parking available as a means of reducing car journeys and increasing use of public transport, walking and cycling. As with road user charging, a vital element in the effectiveness of the policy will be the use made of the proceeds to improve transport choice locally. That expenditure may have to take place in more than one local authority area.

4.108 We propose to legislate to enable the parking charge to apply to all types of private non-residential workplace parking, although we will consult on whether there should be any national exemptions (eg for emergency vehicles and Orange Badge holders). There are strong arguments for workplace parking charges to be levied in all types of location, whether in the town centre or at out of town sites, in order to be consistent with our planning policy, particularly on the revitalisation of towns and cities, by influencing individual's travel choice and businesses' location choice.

4.109 As with congestion charging, subject to the necessary legislation being in place, **we will work with local authorities in developing pilot schemes, individually approved by the Secretary of State** (in Scotland, by the Scottish Executive). The effects will then be monitored so that detailed guidance can be developed for further schemes. We will consult further on the details of how the new workplace parking scheme would operate in practice, the implications for local government finance arrangements and for particular sectors of society, including disabled people. We envisage that Regional Planning Guidance would set out the regional framework within which local authorities would be able to exercise discretion on the specific application of the powers to reflect local circumstances. Local authorities would set out their proposals for use of these powers in the *local transport plan*, showing how a parking charge would support the implementation of their development plan.

4.110 Road user charging and taxation of workplace parking will offer local authorities significant new powers for tackling congestion and pollution in

---

9 A recent Bristol based study of parking control strategies found that a package of measures based on a reduction of 12.5% in private, non-residential parking could reduce future morning peak traffic by between 7% and 10%.

their areas. They will also provide those authorities with significant new sources of revenue for funding improvements, for example in public transport, walking and cycling. Local people, business and other interests must be actively involved to ensure that their proposals attract support.

4.111 We will ensure that schemes are designed and implemented in ways which support the vitality of town and city centres and do not result in dispersal of development. We will not permit their use as a general revenue raising device. We will start with a strong presumption against allowing both new charges to be levied in the same area at the same time. But we will consider proposals where road user charging was applied in one part of an authority's area, and a workplace levy in another.

4.112 We will work closely with local authorities and other organisations to develop guidance on the use of these new powers and to ensure that experience from early pilot schemes informs their application elsewhere. The guidance will cover the needs of disabled people who are car-dependent.

Non workplace parking

4.113 Free parking at other developments (eg for customers and visitors to retail and leisure facilities) also contributes to local congestion, both in town centres and other places. This is particularly so for larger retail and leisure developments, although the effects are not as concentrated in peak hours when compared with commuting journeys. Generous parking provision at such places contributes to low density development, often on the edge of or outside towns, that may not be readily accessible other than by car. For new developments the planning policies now being implemented should ensure that car parking space is limited to the minimum necessary and that full provision is made for public transport access. But more needs to be done for existing developments.

4.114 We have considered whether non workplace private non-residential parking should also be subject to the charge, but have decided that the pilot schemes should be restricted to workplace parking so that the results can be assessed. In the meantime, we propose to tackle over-dependency on the car for access to other types of development in a different way, building on the initiatives which some major retailers have already taken, for example, by improving public transport access to their stores or through home delivery services. We will seek closer partnership between local authorities and all the major retailers/leisure operators in their areas. The types of development to be covered and the measures to be taken would be proposed by the local authority as part of its consultation on the *local transport plan*.

4.115 In preparing the transport plan, local authorities will have to work with retailers and operators of leisure facilities to identify appropriate measures funded by the private sector to reduce car dependency for access to these developments. Such measures should, in particular, help to ensure that people without a car have access to a wider range of goods and services than at present.

4.116 We will be looking for significant progress to be made – especially for larger developments – in the form of better access by public transport, walking and cycling and reduced car dependency for travel to such sites. The measures we envisage are already provided by some retailers and include providing bus shelters and timetable information, funding bus priority measures on the surrounding road network, and providing or supporting bus services to and from the site for customers and staff. Secure pedal cycle parking should be provided as a matter of course. Retail outlets could also extend or introduce easy and affordable home delivery services. These measures would need to be co-ordinated with local authorities' own proposals for improving public transport in their areas so as to maximise the benefits from the contribution of both the public and the private sectors. Local authorities would need to evaluate the impact of such measures on the targets they set in the local transport plan.

4.117 We will ask the *Commission for Integrated Transport* to assess the effectiveness of this approach in meeting the twin aims of reducing the need to travel by car and improving access to goods and services for people without a car.

Sending the right signals

Economic instruments

4.118 The Royal Commission on Environmental Pollution noted that the costs paid by transport users do not reflect the environmental damage and disbenefits caused by the use of land for transport infrastructure and by movements of vehicles. These costs include noise, nuisance and pollution. The failure to take these wider impacts into account can result in misleading signals to users, with consequences for modal choice and travel habits. The use of economic instruments, such as pricing measures and taxation, is an important way of influencing travel choice. Such measures can help to ensure that all costs, including environmental costs, are reflected in the price of transport.

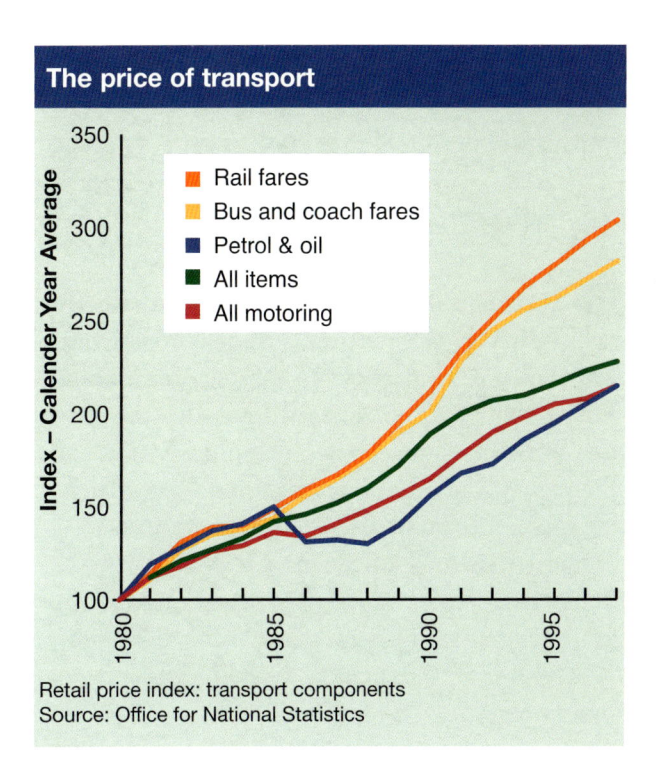

The price of transport

Retail price index: transport components
Source: Office for National Statistics

4.119 We have sent clear signals about the need to use transport more sustainably, consistent with the Chancellor of the Exchequer's 'statement of intent' on using environmental taxes in combination with other instruments to achieve environmental objectives. The March 1998 Budget set out the forward agenda in relation to transport.

4.120 We already use duty on different fuels as a way of influencing demand for road transport and as an incentive to consumers to buy more fuel efficient and less polluting vehicles. We are looking at ways of using taxation on vehicles to achieve similar results. More direct charges can also be very effective in influencing demand for travel or the choice of mode [and we described earlier in this Chapter our plans to enable road user charging and a levy on non-residential parking]. The European Commission's Green Paper "Towards Fair and Efficient Pricing in Transport" advocates the use of such measures to promote sustainable mobility and to enhance competitiveness. The Commission's forthcoming White Paper on transport charging is expected to develop this theme further, with long term proposals for a fair and transparent charging framework for commercial operations across the EU.

CHAPTER 4 — Making It Happen

Cleaner, more fuel efficient vehicles and fuels: fiscal incentives

4.121 **Increasing fuel duty has proved an effective way of directly influencing CO_2 emissions from road transport as part of our strategy for tackling climate change.** It encourages drivers to consider their transport choices when planning journeys and when buying, maintaining and using their vehicles; it also provides an incentive for manufacturers to improve the fuel efficiency of new vehicles. We have adopted a strategy of annual increases in fuel duty of at least 6% on average above inflation, 1% higher than the previous Government's commitment.

4.122 **We will continue to encourage the use of more environmentally friendly fuels.** A key to increased use will be industry responding to our lead by much more widespread provision at filling stations. In March 1998, we increased the duty differential between ordinary diesel and ultra low sulphur diesel (ULSD) from 1p a litre to 2p a litre and next year we intend to increase it to 3p a litre. We have also tightened the specification of ultra low sulphur diesel to confine it to the cleanest diesel fuels. Our aim is to reduce emissions of particulates and nitrogen oxides from existing vehicles and over time to encourage the use of cleaner diesel technology. This is an essential element in our strategy to improve air quality, particularly in urban areas. In addition, we are moving towards a fairer treatment of diesel and petrol, based on the energy and carbon content of these fuels. This will mean that the duty on diesel should be higher than on petrol. We began this process in March 1998: duty on diesel is now 1p a litre more than unleaded petrol, and we intend to increase this differential in future years.

> **London Transport Buses**
>
> - carried out an extensive review of options for reducing emissions and now have a policy of switching all their bus fleet to cleaner diesel such as City diesel;
>
> - around half of their London services use cleaner diesel and they are working towards 100% use on all routes;
>
> - oxidisation catalysts have been fitted to 950 buses and a trial of 20 buses with particulate traps is underway;
>
> - we welcome these initiatives and hope to work with LT Buses and operators to build on these developments.

4.123 We have frozen the duty on road fuel gases, following significant reductions in previous Budgets, reflecting the fact that they produce much lower emissions, especially of particulates, than diesel. In doing so, we have widened the differential between these fuels and diesel and we are committed to at least maintaining the differential that existed in July 1997 in order to provide incentives and certainty for potential investors in these new technologies and cleaner fuels. In addition, from April 1999, the cost of conversion of company cars to gas power will no longer be included in the tax calculation of employee benefits.

Government Car Service (GCS) – leading the way

- GCS decided last year to convert as many as possible of its 150 cars to run on alternative fuels – either Liquified Petroleum Gas (LPG) or Compressed Natural Gas (CNG) over a 5 year period;

- the decision was taken both on environmental and cost grounds;

- the newest CNG and LPG vehicles offer lower emissions of regulated pollutants than diesel and petrol and it is cheaper to convert all government cars;

- GCS will be closely monitoring the environmental performance of vehicles with a programme of emissions tests;

- this initiative will be important in stimulating the market for gas fuelled vehicles and reducing pollution.

Improving air quality through technology: remote control barriers ensure priority for this gas-powered bus in Cambridge

4.124 **We have introduced legislation in the 1998 Finance Bill which will provide an incentive for cleaner vehicles through the vehicle excise duty arrangements (VED) for lorries and buses.** From January 1999, lorries and buses producing very low particulate emissions will receive an incentive of up to £500 off VED rates to encourage owners to achieve tough emissions standards: for example, by fitting particulate traps to vehicles, fitting higher standard engines or switching to road gas fuels.

4.125 **We will extend our reform of vehicle excise duty by introducing a new system of graduated VED for cars from next year, which will include a new lower rate of £100 for the smallest and least polluting cars.** We will consult on the details shortly. Our aim is to encourage motorists to take account of environmental impact when buying new or second hand cars. In the meantime, we have frozen VED for cars and lorries, representing a fall in real terms, and helping to shift the burden of taxation from car ownership to car use. The freeze on VED for lorries should also help maintain the competitive position of UK hauliers and partly offset the impact of higher fuel duty.

4.126 The Chancellor has already announced that he will **review the system for setting VED rates for lorries to ensure that the environmental damage they cause is reflected in their VED rates.** This review will take into account the wider environmental impacts of lorries as well as their physical effects on the road infrastructure. The present system does not, for example, take axle loading, which has a substantial effect on road wear, properly into account. With the introduction of new lorry weights (see Chapter 3) it will be particularly important to ensure that those operating vehicles with 11.5 tonne maximum axle weight pay a rate of VED commensurate with their increased road costs in comparison to vehicles operating at a maximum of 10.5 tonnes. This review will therefore consider the VED rates for these new lorry weights.

4.127 Local bus services have benefited from a scheme for Fuel Duty Rebate (FDR) since 1962, designed to avoid increases in fuel duty feeding through into fares. This was worth some £230 million annually to local bus services. **This year, for the first time since 1993, we are increasing the rebate by £40 million in line with the duty increase on diesel, to protect bus operators from this year's fuel duty increase and ensure that the cost of bus travel rises by less than the cost of motoring.** We will consult shortly on our plans to target Fuel Duty Rebate on rural services and on more environmentally-friendly vehicles.

Company cars

4.128 Company cars account for almost 20% of car mileage and over half of new cars are first registered in a company name. Company policy on the purchase and use of company car fleets is therefore important for the environment. Company cars are generally much newer and better maintained than the average private car and therefore less polluting per unit of fuel consumed. However, they tend to have larger engine sizes than the average private car and as they account for a high proportion of the new vehicle fleet, they contribute to higher overall average fuel consumption both directly and through their influence on the stock of cars in the second hand market. Around 1.65 million company cars are available for private use. These drivers also tend to drive significantly further to and from work and those who receive free fuel drive further still.

4.129 We recognise that some drivers have to use a car because of the nature of their work. However, the existing system for taxing company cars has been criticised as providing a perverse incentive to drive further in order to reach the business mileage thresholds which attract significant reductions in the tax liability. In the March 1998 Budget, the **Chancellor announced that he would be considering the case for replacing the existing business mileage discounts with discounts for driving fewer private miles in company cars**, and invited people to send comments to the Inland Revenue. So far, a wide range of organisations and individuals have responded.

4.130 The current tax system for employees who receive free fuel from their employers for private use (about half of all company car drivers) has given them little incentive to reduce their private mileage, as the employee pays the same amount of tax whatever the amount of private mileage driven. It is important to send consistent messages about the need to reduce unnecessary journeys and improve fuel efficiency. **We therefore announced in the March 1998 Budget that we will increase the scale charges for employees provided with free fuel for private use by 20% each year over and above normal increases up to 2002/3 to discourage employers from providing free fuel.**

Incentives for green travel

4.131 We recognise the interest that has been shown in the role of tax incentives in encouraging more environmentally friendly forms of travel, particularly in supporting green transport plans by employers for their staff – for example, through new tax incentives for public transport season tickets. At present interest free loans of up to £5,000 provided by employers for the purchase of season tickets are not taxed as a benefit in kind. An important issue is how employers and employees would respond to further changes in the tax system, and the relative impact of tax measures compared with other factors. **We therefore intend to undertake research on the importance of such factors, including the influence of the existing tax system with a view to seeing whether changes could be effective in promoting green transport plans.**

4.132 Not all employers may be aware of the possibility, under the present tax system, of offsetting against taxable income, expenditure they incur in encouraging their employees to use green transport – such as the running expenses of a bus or coach to bring employees to work. Capital allowances may be available for capital expenditure. But where such benefits are provided, employees may face an income tax charge. The Inland Revenue intend to publish further guidance on the tax rules later this year, and in the meantime employers should check the position with their tax offices at an early stage in their planning.

Setting standards

4.133 Setting higher environmental standards for vehicles and fuel has made a significant contribution to reducing the harmful effects of emissions from road traffic. Further improvements at the EU level will be necessary to achieve our objectives for reducing greenhouse gas emissions and improving air quality. We will continue to work closely with the European Commission and other Member States to secure this.

Cleaner, more efficient vehicles and fuels: standards

4.134 The European Auto-Oil programme provides for improvements in vehicle technology and fuel quality through higher standards. **Tighter emission standards for new cars and light vans will apply from 2001. These will be 20-50% more stringent than those currently in force.** Technical improvements to maintain the emissions performance of vehicles throughout their life will be required from the same date.

4.135 Improvements in the design of lorries have resulted in gradually improving energy efficiency with lower pollution emissions. At the European level, the introduction of emission standards for new lorries in 1991, 1992 and 1996, together with the proposed standards for 2000 will reduce emissions still further. But in the longer term even with the introduction of these standards and those for lighter vehicles, continuing growth in road traffic could start to erode the benefits in terms of air quality.

4.136 Auto-Oil will also lead to a further tightening of emission standards for 2006 and beyond. This may represent a reduction of up to 50% on the 2001 standards. The position for lorries and buses is similar, although the new controls will be applied slightly later.

4.137 Under the Auto-Oil programme, cleaner fuels will be required from 2000. These fuels will assist in the achievement of the new vehicle standards as well as reducing emissions from the existing fleet. Significant improvement in fuel quality will also be made in 2005. These will allow the use of the clean vehicle technology required to meet the 2005 vehicle standards.

4.138 We support the European Commission's strategy which aims to reduce emissions of CO_2 from new cars to an average of 120 grammes per kilometre by no later than 2010. This represents an improvement of about a third on the current average. The strategy was agreed by the EU Council of Ministers in June 1996 and as part of this strategy, the Commission is negotiating a voluntary agreement with European car manufacturers to improve fuel consumption.

4.139 We consider that more can be done to reduce the environmental impact of freight distribution and we will support continued research into more efficient vehicle design and the use of alternative fuels.

Better air quality

4.140 Our policy for improving air quality is set out in the National Air Quality Strategy. This sets air quality objectives derived from health based standards for eight pollutants to be achieved by the year 2005. **We are currently reviewing the strategy and aim to produce conclusions by the end of 1998.**

4.141 The concentration of air quality problems in busy urban areas with high levels of road traffic means that national measures to improve vehicle standards and fuel quality will often have to be complemented by local action.

4.142 A key tool for delivering the National Air Quality Strategy is the new system of Local Air Quality Management. Local authorities have a duty to assess air quality in their areas to determine whether the objectives set out in the Strategy, and prescribed in the Air Quality Regulations 1997, are likely to be met by 2005. Where a local authority considers that one or more of the objectives is not likely to be met, as a result of national measures alone, it must declare an air quality management area, covering the area where the problem is expected.

4.143 For each air quality management area, an action plan must be drawn up by the local authority, in consultation with the public and with organisations such as the local health authority, the Highways Agency, the Environment Agency and the Scottish Environment Protection Agency. The purpose of the action plan is to identify measures which will help to achieve the air quality objectives for the area and a timetable for their implementation.

4.144 Many air quality management areas are likely to be places where most of the pollution comes from road transport. Proposals to reduce emissions from traffic will therefore feature in the majority of air quality action plans. Many of these proposals will also relieve congestion and noise, help local authorities to meet road traffic reduction targets and will therefore form part of the *local transport plan*.

4.145 Air pollution does not respect national boundaries, action at the European level is therefore also important. The framework for improving air quality across Europe is set out in a Directive, which provides for limit values to be agreed for twelve pollutants. During our Presidency of the EU, we have secured a landmark agreement on legally binding limit values for three pollutants to which road transport is an important contributor – nitrogen dioxide, particles and lead[10]. This agreement will bring real benefits to the people of Europe, ensuring a high level of protection for people who are particularly vulnerable to air pollution.

Ports and shipping

4.146 Shipping operations can have an impact on the marine and coastal environment. We are committed to working through the International Maritime Organisation (IMO) for the adoption of rules and standards and for their effective enforcement.

4.147 In the UK, a number of routing measures have been established around the coastline; these were introduced to prevent accidents but some also provide protection for environmentally sensitive areas, in the light of the Donaldson report[11]. The effectiveness of these measures and the need for complementary action is kept under review.

---

10 agreement was also reached on limit values for sulphur dioxide. The four agreed limit values are broadly equivalent to the comparable objectives in the UK's National Air Quality Strategy.

11 "Safer Ships, Cleaner Seas", Report of Lord Donaldson's inquiry into the prevention of pollution from merchant shipping, May 1994.

4.148 When maritime accidents occur, it is essential that prompt response action can be taken. That is why we have taken measures to secure the provision of emergency towing vessels (ETVs) at three strategic locations around the UK coast for the next three winters. Ideally, however, we would like to provide all-year round protection. We have therefore announced that, during the course of the current ETV contracts, **we will be considering the scope for imposing a levy on ships to pay for additional ETV cover.**

4.149 We also want to reduce operational pollution from ships by encouraging the responsible discharge of ships' wastes in port through our policy of port waste management planning. The operators of ports, harbours, terminals and marinas have a statutory duty to ensure the provision of adequate reception facilities for ships' wastes. We have complemented this with a new duty to plan the provision of such facilities in consultation with users and other interested parties. The aim is to integrate shipping and port operations so that ships' wastes can be efficiently discharged in port. This should remove any excuse for illegal discharges at sea.

4.150 In the light of the Marine Accident Investigation Branch report on the Sea Empress, a review of the arrangements for harbour pilotage was carried out. We have concluded that habour pilotage should continue to be provided by harbour authorities but increasingly integrated into their overall safety systems. These systems need to be reviewed as a whole if the highest safety standards are to be achieved. **We propose to develop a "Marine Operations Code for Ports" covering all port safety functions, not just pilotage. We are now taking this forward in consultation with industry and other interests.** The aim is to set a national standard, and to create a guide for harbour authorities to prepare detailed safety policies, in consultation with local users and other interests. The Code will also serve as a benchmark by which safety improvements can be measured in future.

Marine clean-up

4.151 We are pressing within IMO for continued international action to ensure that prompt and adequate compensation is available for the costs of clean up activities and losses caused by marine pollution from ships. We welcome the fact that the amounts available under the international compensation regime for oil pollution from tankers were recently increased. We are participating in discussions on further changes to the regime.

4.152 We are taking a leading role in discussions aimed at encouraging the early ratification of the IMO's Hazardous and Noxious Substances Convention. Once in force, this convention will extend the benefits of the existing regime for oil pollution from tankers to other dangerous and polluting cargoes carried by ships. We are also in the lead in IMO discussions on how to fill the gaps left by these two regimes. In particular, we are pressing for the development of an international liability regime for oil pollution caused by the fuel carried by ships other than tankers and of an IMO code setting minimum standards for the insurance cover taken out by shipowners.

Air transport

4.153 **With our international partners, we will continue to press for tighter worldwide standards for aircraft noise and emissions.** We will also ensure that appropriate noise standards are set and enforced at UK airports (see later in this Chapter) and environmental issues will be an important part of our review of airports policy.

4.154 We will seek international agreement to improve the aircraft engine standard for emissions of nitrogen oxides, to reflect what is achievable using the latest technology. We will maintain pressure in the International Civil Aviation Organisation (ICAO) for greater stringency in aircraft noise standards to reflect advances in technology.

We will also seek agreement to prevent the use of aircraft which barely meet existing noise standards. Where worldwide measures on engine emissions and aircraft noise cannot be agreed, we will consider what can be done at European or UK level.

4.155 The Kyoto climate change Protocol places a new obligation on countries to work through ICAO to limit or reduce emissions of greenhouse gases from aviation. **We will continue to pursue in ICAO the potential for environmental levies and to press for removal of the exemption from tax on aviation fuel, to encourage fuel efficiency.** We welcome the initiative by the previous Dutch Presidency of the EU for a study on the competitive and environmental effects on the EU of an aviation fuel tax, and look forward to its publication later this year.

Better planning

4.156 Land use planning plays a central role in delivering sustainable development, complementing and contributing to the success of other measures such as economic instruments. The planning system operates by providing incentives for the development of land through the allocation of uses in statutory plans and a means of control, by preventing development which is judged to go against the public interest.

4.157 **By reviewing the framework set in planning policy guidance and by ensuring that Regional Planning Guidance is up-to-date and incorporates regional transport strategies, we can help to ensure that local authorities' plans and decisions, and proposals from individuals and businesses, reflect integrated transport policy.**

4.158 Our overall approach to planning is aimed at containing the dispersal of development so reducing the need to travel and improving access to jobs, leisure and services. We want to promote regional strategies for planning that are integrated and sustainable and we want these to provide the context for *local transport plans* and development plans.

4.159 We will monitor the impacts of our planning policies to ensure that they are well targeted and do not impose unnecessary costs on business. This is central to our initiatives to modernise the planning system to improve its efficiency and effectiveness.

New policy guidance

4.160 The publication of PPG13 (the planning policy guidance note on transport) in England in 1994 was a major step towards planning land uses and transport together. It aimed to reduce the need to travel, especially by car, and to encourage means of travel which are more environmentally friendly.

4.161 We will build on this change of direction and, based on a clear assessment of the costs and benefits, **we will update our planning guidance to ensure that we have the right framework to deliver integrated transport policy at the local level. In England, we will revise our guidance notes on Transport, Development Plans and Housing.**

4.162 In Scotland, we are publishing consultative drafts of a National Policy Guideline and a Planning Advice Note on *Transport and Planning*. These documents will take forward in Scotland the general principles outlined below on the integration of land use and transport planning. The guidance provided in Planning Guidance (Wales): Planning Policy, Planning Guidance (Wales): Unitary Development Plans, and appropriate Technical Advice Notes will be revised to set the planning framework for applying the new approach. In Northern Ireland the new policy framework is set out in the Regional Strategic Framework and planning policy statements will be prepared or revised.

Planning guidance for transport

4.163 **We will update existing guidance on locations for major growth and travel generating uses, with an increased emphasis on accessibility to jobs, leisure and services by foot, bicycle and public transport.** This will include the promotion of major development within public transport corridors and other areas where good public transport exists or can be provided. We have research in hand to provide practical advice for local authorities so that their proposals for growth along public transport corridors are brought forward in ways that support sustainable development.

4.164 We will ensure that development plan policies for parking support our policies for the location of development. Parking standards should be devised and applied having regard to the accessibility of locations by modes other than the car. We have commissioned research to help in developing a method for local authorities to set parking standards, by type and location of development. This will report by the end of the year. We will also encourage cycle parking standards to be applied more widely.

4.165 **Development plans should give better protection to those sites and routes (both existing and potential) which could be critical in developing infrastructure to widen transport choices;** such as interchange facilities allowing road to rail transfer or for water transport. Alternative uses related to sustainable transport should be considered first for sites now surplus to transport requirements. More generally, before giving permission to new developments, local authorities should consider carefully the effect on sustainable transport objectives.

4.166 **We will provide further guidance on how land use planning can promote public transport, cycling and walking and clarify the handling of traffic management issues in development plans.** We will set out guidance on the land use issues arising from re-allocating road space to pedestrians, cyclists and buses. There will be new guidance to ensure that planning policies and implementation should take full account of the needs of all in society, including those of disabled people and the need in rural areas to promote service provision.

Housing

4.167 We will revise our planning guidance on housing to give clearer advice on the location and form of housing development. This will emphasise the benefits of providing new homes in towns and cities and making the most of places or vacant buildings which can be well served by public transport or easily reached on foot or by cycle. Local planning authorities will need to consider the future travel patterns that would be created when planning for new homes.

4.168 The revised guidance will stress the need for careful planning of those places and sites that are not close to existing public transport. Our aim is for new housing which avoids undue reliance on the car. The options available to local authorities will include ensuring that any major new development provides good public transport as part of the scheme, or where this is not feasible using the place for activities that do not generate significant travel demands.

Chapter 4 — Making It Happen

Development plans

4.169 We consulted earlier this year on our proposals to improve the procedures for preparing development plans and their content. We will publish shortly a full revision of the guidance as a draft for public consultation. This will:

- set out the new approach for producing better plans more quickly;
- provide guidance on how development plans will integrate with *local transport plans*.

4.170 In revising planning guidance on minerals we will take account of the importance of promoting greater use of rail and water transport.

Better implementation in the planning process

4.171 Revising planning guidance, moving to a new generation of Regional Planning Guidance and encouraging better appraisal of plans and proposals will not in themselves produce developments that support the objectives of this White Paper, without the co-operation and support of the development industry and local authorities.

4.172 We are confident of winning that support in most cases. But where there are proposals that would undermine our integrated transport policy, we will not hesitate to consider using our powers of intervention, through the scrutiny of plans by Government Offices in England and through decisions by the Secretary of State on applications before him.

> **Informing key decision-makers**
>
> - in partnership with the professional bodies, we will develop a strategy for training which brings key players and decision-makers together and raises awareness;
> - this will promote greater understanding of integrated transport policy and its implications for town and country planning;
> - we will explore the scope for a joint initiative with the Local Government Association and the Royal Town Planning Institute.

4.173 We have announced measures to modernise the planning system in England. These include more explicit national policy statements on the need for projects of national importance such as airports and the criteria for site selection, and changes in public inquiry procedures.

4.174 We will continue to look for ways to improve the planning system's delivery of integrated transport policy. This will include a review of the Use Classes Order and General Permitted Development Order. We will also update the guidance on the use of planning conditions to clarify the scope for developers to provide facilities for pedestrians, cyclists and those travelling to and from the new development by public transport.

4.175 Similarly, once our review of the use of planning obligations is completed, we will use the opportunity of any subsequent revision to guidance on their use to shift the emphasis when improving off-site transport facilities away from catering for car traffic to providing for public transport and cycling and walking. We will also consider encouraging the incorporation of green transport plans into planning obligations.

Good design

4.176 Good design of new development is important if we are to make the most of opportunities for walking, cycling and public transport in new developments. We are publishing two documents which stress the importance of design issues in the development process: the "Good Practice Guide on Design in the Planning System" sets out key urban design principles, including ease of movement and mixed use development, to maximise the opportunities for public transport, cycling and walking. "Places, Streets and Movement", a companion guide to Design Bulletin 32, advocates distinctive, safe and attractive estate layouts which better reflect local character, and moves away from planning which caters solely for car use.

Better enforcement

4.177 Many of the proposals in this White Paper will require new measures to ensure that they are properly enforced. Although we are confident of general public support for these measures, it will be eroded if some people are seen to ignore the standards that are necessary for the good of us all.

Better enforcement: road traffic

4.178 Previous governments have, we believe, done too little to combat the idea that many road traffic offences – speeding in particular – do not really matter. We also want better enforcement to give priority to public transport and to reduce pollution from traffic. So we need to take a new and radical look at enforcement.

Strengthening enforcement: the Vehicle Inspectorate check emissions.

4.179 Better enforcement does not necessarily require more resources. Nor is it necessarily right that all these tasks should be undertaken by fully trained police officers. Clearly there are some motoring offences which amount to very serious crimes. We want these to be detected and prosecuted with all the powers which the police possess. But we also want to make full use of other means, especially technology, to make the most of the police resources we have and to assist other agencies to enforce safety and environmental standards effectively.

Technology for enforcement

4.180 In recent years, cameras have been used as a very effective means of detecting speeding or jumping red traffic lights. A Home Office study in 1996 showed that speed cameras operated by 10 of the 43 police forces in England and Wales reduced accidents by 28%. The investment paid off – there was a 500% return in the first year and a similar level of benefit thereafter. We want to see an extension of the use of cameras to reduce speed and accidents, save lives and improve the environment.

4.181 As part of our commitment to greater efficiency and effectiveness in public services we are looking at ways to improve the current arrangements for funding cameras. In doing so, we will want to ensure that the use of cameras is determined by the road safety and other benefits which they can deliver for the community.

4.182 Recently, more London buses have been equipped with cameras so that car drivers who use or obstruct bus lanes can be caught and prosecuted. Illegal use of bus lanes carries a maximum fine of £1,000 but many offences are dealt with under the fixed penalty procedure, in which case the penalty is £20. **We take a serious view of bus lane offences and will consider the case for a higher penalty in our review of the level of fixed penalties.** Trials have also been carried out using cameras to enforce box junctions where offences can cause congestion. Many of our proposals will require better traffic management and we will investigate what further contribution cameras might make to enforcing such measures.

4.183 In the longer term, it may be possible to use adaptive cruise control devices linked to geographical information systems to prevent vehicles from going faster than the speed limit in specific places. The technology is still under development but there are also potential future applications for congestion charging. Other future developments include devices which would maintain safe distances between vehicles and ways of slowing traffic automatically when approaching traffic lights or at junctions or outside schools.

Role of other agencies

4.184 We have traditionally looked to the police for most forms of enforcement. But more specialised staff can fulfil some of these tasks, leaving the police to do more difficult ones. For example, there has been a move away from the use of police for parking enforcement to traffic wardens, some under police control, others employed by local authorities. In other areas, police duties have been civilianised – for example, photographs taken by speed cameras are commonly examined by civilians supervised by police officers.

London bus lane enforcement

- a publicity campaign will highlight the importance of bus lanes in improving the speed and reliability of buses and will show the difference it will make when drivers obey the law. The aim is to improve journey times and reliability, encouraging greater use of buses;

- police and local authorities will work together so that restrictions will be enforced on a route by route basis, concentrating on bus lanes, bus stops and key junctions or areas where bus priority measures would be particularly helpful;

- we will be rolling out camera enforcement of bus lanes London-wide, harnessing new technology to help keep bus lanes clear;

- London Transport is to offer London boroughs' parking attendants free travel passes, allowing them to ride on buses and target illegal parking in bus lanes.

Greener Cities – local authority emissions checks

- seven local authorities (Birmingham, Bristol, Canterbury, Glasgow, Middlesbrough, Swansea and Westminster) are piloting powers to check vehicle emissions at the roadside, with the police, and issuing fines to owners of vehicles which fail the test;

- the results are being monitored. If the scheme is successful, the powers will be extended to all local authorities;

- to help motorists, free emissions checks are on offer in the pilot areas.

4.185 Some of the tasks needed to make our new policies a success are relatively routine; they do not require the skill and judgement of a police officer. For example, it should not always be necessary, provided there are safeguards, to employ the police to stop traffic so that other specialists can check vehicle emissions or the proper loading or roadworthiness of lorries. Cameras can help in some of these tasks by making it unnecessary to stop vehicles, but not all problems can be dealt with that way. **We will consult the police associations and others on proposals that:**

- traffic could be stopped by traffic wardens under the control of the police for checking by other agencies, with essential safeguards;

- an enforcement police sign to stop traffic, put in place by a police officer, could be used by agencies such as the Vehicle Inspectorate to stop lorries, buses and coaches at checkpoints for roadworthiness and compliance with other traffic legislation;

- police civilian staff should be able to take decisions about camera offences within criteria set by the chief officer. As at present, the criteria would reserve particularly serious offences for prosecution in the courts, and maintain overall control over fixed penalties.

4.186 We will examine whether other less serious offences where proof is relatively easy to obtain might be suitable for such streamlined procedures, to save police and court time. In taking forward these initiatives, we will ensure that traffic rules and regulations are enforced in a way which respects the rights of individuals, and is properly targeted on bad driving and anti-social behaviour.

Police organisation

4.187 Chief officers of police are responsible for the level of policing in their areas. They have to have regard in applying their resources to the key objectives set every year by the Home Secretary. Although road policing is not one of the Home Secretary's key objectives, these are intended to indicate development priorities for the coming year and, as the Home Secretary's letter to chief officers on the key objectives made clear, "traffic policing is a central part of the police's responsibilities for maintaining law and order and preventing and deterring crime and reducing death and injury on the roads. I will therefore expect traffic policing to play a full part in achieving my overall objectives for the Police Service, particularly in relation to community safety and crime reduction and in achieving a safer environment on the roads". We will continue to consider ways to promote effective road policing.

British Transport Police

4.188 The British Transport Police (BTP) provides a specialist dedicated police service to the railway industry in Great Britain, covering railways, London Underground and the Docklands Light Railway. The BTP's aim is to make railways safe and secure and it is committed to preventing, detecting and reducing the fear of crime. This includes a significant role in counter-terrorism, to protect the public and the railway system. The annual budget of some £104 million is provided by the industry, primarily train operators and Railtrack.

4.189 We are reviewing the future of the BTP, including the jurisdiction of the force and the merits of establishing a separate police authority to increase its public accountability. We expect to announce conclusions shortly.

Wheelclamping on private land

4.190 The law on the use of wheelclamping on private land is clear, but largely unenforced. In

England and Wales wheelclamping on private land is legal providing there are warning notes and the release fee is reasonable. Unscrupulous wheelclamping operators are preying on motorists and cause nuisance to the public. In response to our recent consultation, there was overwhelming support for regulation of wheelclamping on private land.

4.191 We wish to introduce regulation of wheelclamping in the context of the introduction of statutory measures to regulate the private security industry as a whole. We intend to bring forward firm proposals for regulating the industry later this year.

Better enforcement: freight transport

4.192 We will bring forward a number of initiatives to promote higher safety and environmental standards in the freight transport industry. Effective enforcement here often depends on reliable information. At the moment, this information is spread among different enforcement organisations.

We will:

- **improve the co-ordination** of information between Government and other agencies involved in enforcement activities (ie the police, the Vehicle Inspectorate, the Driver and Vehicle Licensing Agency, the Traffic Area offices and local authority Trading Standards officers) using information technology through the Joint Enforcement Database Initiative.

A significant minority of transport operators continue to adopt a short-term, cost-cutting approach at the expense of safety and the environment. **We will:**

- **bring forward legislation to enable the detention of illegally operated vehicles,** in the light of the current consultation;

- seek to ensure that weighbridges are available for enforcement at all freight terminals and ports where this is justified by the levels of road freight traffic.

We will also work in partnership with industry and local government to promote a more responsible attitude to safety and environment throughout the freight transport industry. As part of this, **we will:**

- **promote best practice on safety and the environment with industry and the larger logistics companies.**

Better appraisal

Transport impact assessment

4.193 Transport impact assessments will be incorporated in the process of assessing the environmental implications of all relevant Government policies and major location decisions. Arrangements are also being made for Health Impact Assessments of key policies. This will ensure that transport, environment and health implications are considered in Government policy making where appropriate.

> Building in transport and land use considerations in the early stages of policy development, Government leading by example:
>
> - in DOH's Green Paper "Our Healthier Nation";
> - in DfEE's "The School's Environmental Assessment Method";
> - in MoD's guidelines for disposing of redundant sites;
> - in the adoption of green transport plans.

Better appraisal

4.194 We are developing a checklist of women's transport requirements which can be used to audit transport initiatives and ensure that their transport needs are taken into account from the start. We will make this checklist available to all local authorities and transport providers.

New approach to appraisal for transport projects

4.195 We are developing a new approach to the appraisal of different solutions to transport problems[12]. This is designed to draw together the large amount of information collected as part of the appraisal of a transport problem and alternative solutions. This information is set against the five criteria which we have adopted for the review of trunk roads ie integration, safety, economy, environment and accessibility. It looks at the contribution of different forms of transport in developing alternative solutions and the potential effect of the new integrated transport approach, including the scope for and effect of demand management measures. It is our intention that this new approach, once finalised, will be applied to the appraisal of all transport projects, including proposals for all road schemes.

Economic appraisal

4.196 As part of this appraisal, the economic impact of road investment is taken into account largely through the estimated benefits of reduced journey times for commercial, business and other traffic. Techniques are being developed to assess the economic value of journey time reliability and to identify those improvements which contribute most to reliability. In the interim, the current review has sought to identify such benefits in qualitative terms.

4.197 The Standing Advisory Committee on Trunk Road Assessment (SACTRA) is considering the relationship between transport infrastructure investment and economic growth. Its interim report noted that transport investment can have economic impacts that are not measured in conventional cost-benefit appraisal and which could be either positive or negative. SACTRA is undertaking further work to determine the feasibility of developing appraisal methods to assess these additional impacts.

4.198 We accept SACTRA's interim finding that there is no simple, unambiguous link between transport provision and local regeneration. Each case must be assessed on its merits. In some cases, road access is essential in order to provide access to sites which could not otherwise be developed. We agree that improvements are needed to the methods used in local and regional economic impact studies so that the contribution of transport investment to regeneration can be better assessed. We will take account of SACTRA's interim findings in developing our new approach to appraisal. SACTRA is now continuing its investigation and expects to report at the end of the year.

Environmental appraisal

4.199 Environmental impacts are taken into account from the earliest stages of planning and designing new transport projects. Environmental appraisal considers a range of effects including air quality, noise, emissions, land, wildlife, the countryside, the built environment and cultural heritage as well as the effects on people and their health. Because of the potential environmental impacts of major new construction, it is important that alternatives to new construction are considered at the earliest stages of planning. Alternatives include making better use of existing infrastructure and managing demand for it and the use of other forms of transport.

---

12 Appraisal frameworks based on these principles will be developed for assessing transport proposals in Scotland, Wales and Northern Ireland.

4.200 All projects likely to have a significant effect on the environment are subject to a formal environmental impact assessment in accordance with EU legislation. Proposals for transport infrastructure affecting sites of international importance are assessed in accordance with the provisions of the relevant international legislation. In practice, we expect there to be few cases where it is judged that imperative reasons of overriding public interest will allow development to proceed which will have an adverse impact on the integrity of internationally designated sites.

4.201 For all environmentally sensitive areas or sites[13] there will be a strong presumption against new or expanded transport infrastructure which would significantly affect such sites or important species, habitats or landscapes. Where such proposals arise, they will be assessed in relation to the status and purpose of the site including whether it is of international, national or local importance, and where relevant, the protected status of the species or habitat in international or domestic legislation and whether it is a target species or habitat in the UK Biodiversity Action Plan, against the degree of impact of the scheme and the scope for mitigation.

4.202 A transport scheme which would significantly affect a sensitive site or important species, habitat or landscape should not go ahead unless it is clear that the net benefits in terms of the other objectives (including other environmental benefits) clearly override the environmental disbenefits, there is no other better option and all reasonable steps have been taken to mitigate the impact. Particular consideration should be given to species or habitats given international protection, for example, under the EC Birds and Habitats Directive. Each case will be determined on its merits, taking account of the following questions:

- how important is the area/site, including its international importance or significance for UK biodiversity?
- how serious is the likely impact?
- are there alternatives which avoid the impact (including not going ahead with the scheme)?
- would the alternatives serve the purpose and at reasonable cost?
- if not, are mitigation or compensatory measures feasible? Are they likely to be successful? Are the costs reasonable in the circumstances?

4.203 The feasibility, desirability and cost of providing compensatory measures will be a factor: some areas, sites, habitats or species may be irreplaceable and that will have a particular weight in the assessment. These principles will be applied to all forms of transport development which affect sensitive areas or significant aspects of our cultural heritage, such as listed sites or buildings.

4.204 In order to improve the appraisal of environmental impacts, the statutory environmental agencies are developing the concept of environmental capital. We will keep in close touch with this work and incorporate its findings as appropriate in our assessment and appraisal of new infrastructure.

Improving appraisal: the planning process

4.205 Research shows that the former DOE publication, "The Environmental Appraisal of Development Plans: a good practice guide"[14] has been widely adopted and is proving useful.

4.206 Our emphasis now goes wider than environmental impacts, encompassing the full agenda of sustainable development, including

---

13 For example, Sites of Special Scientific Interest, National Parks, Areas of Outstanding Natural Beauty, National Nature Reserves and National Scenic Areas in Scotland.

14 "The Environmental Appraisal of Development Plans: A Good Practice Guide", DOE, 1993. HMSO ISBN 0-11-752866-8.

tackling social exclusion. We will therefore consider complementing this guide with one which considers social, economic and environmental impacts, so that decisions can be taken with the benefit of a consistent appraisal. This would help local authorities to improve their understanding of the likely economic and social impacts of emerging policies and proposals in draft development plans.

4.207 We have research in hand to assess the practicability of applying the principles of such appraisal to Regional Planning Guidance, drawing on experience so far in the UK and elsewhere.

4.208 We will consider how we can improve the way local road proposals and transport issues are handled at structure plan Examinations in Public (EIPs) and at local plan inquiries. We are considering in particular the adequacy of the appraisal methodologies used by local authorities and the thoroughness of the evaluation undertaken, as well as the time allowed for transport issues at EIPs and inquiries and the extent to which these issues are dealt with in the resulting reports.

Improving appraisal: development proposals

4.209 Properly considered development plans are important but achieving better integration on the ground depends also on getting the right development in the right places.

4.210 Currently, some proposals for major development are subject to 'environmental assessment' and 'traffic impact studies'. Our planning guidance on transport (PPG13) and on Retail Development and Town Centres (PPG6) both advocate a broader approach to appraisal to encompass the impact on overall travel and car use. We will provide further guidance on how development proposals should be assessed, including accessibility to the site by public transport, walking and cycling.

4.211 These documents support our planning policy guidance that places increased emphasis on the need for local authorities in preparing development plans to consider design. Local authorities can help by promoting good design and rejecting bad design. Applicants for planning permission will be expected to demonstrate that their proposals have addressed key design principles.

Planning for accessibility

We are considering ways of giving greater emphasis to accessibility, in the sense of access to jobs, leisure and services by public transport, walking and cycling, in the land use planning process. These include reinforcing our policies for the location of development with accessibility criteria at the regional and local level, broadening the approach to assessment for development plans and proposals, and in determining parking standards.

Understanding the effects of noise

4.212 We have work in hand to develop a fuller understanding of the problems arising from transport noise. This includes:

- a study to measure noise at a representative sample of 1,000 sites in England and Wales with results available in 1999-2000;

- an assessment of attitudes to noise on the same timescale (we are exploring the practicability of linking these two surveys as this might allow correlations to be drawn between noise levels and attitude to noise);

- work with Birmingham City Council to carry

out detailed noise mapping of its area, to be completed in Autumn 1998 and possible follow up work. The aim is to test the usefulness of noise mapping as a tool for establishing where noise problems exist and for assessing the noise effects of proposals for traffic management;

- further work on the health and related effects of noise will be carried out under a three year joint Department of Health and Department of the Environment, Transport and the Regions research programme.

Noise standards

4.213 At the same time, we will continue to apply measures to reduce noise and the impact of noise from transport. The European Union has a role to play in developing standards for vehicle emissions and we will press for the adoption of standards aimed at reducing noise emissions where this can be done cost effectively and without jeopardising safety. We will work with the European Commission on the development of noise standards for new freight wagons and high speed inter-city trains.

4.214 The adoption of tighter European noise emission standards for road vehicles over the last ten years has had a noticeable effect on noise emitted from vehicles in urban areas, but not so far on the noise from traffic travelling at higher speeds on inter-urban roads. Although the noise emitted by vehicles built to the latest standards is about half that allowed ten years ago, the benefits are being eroded by the continuing growth in traffic. A new standard has been developed specifically to limit tyre noise, but attention is increasingly being directed at road surfaces which generate less tyre noise.

4.215 The noise arising from the latest quiet road surfaces compared to that generated by the traditional types of motorway surface is about the same as if the amount of traffic were halved. We are continuing to develop quieter surfaces to improve their noise reduction properties, extend their useful life and reduce costs. In doing so, we are taking particular care to maintain a sufficient degree of grip between road and tyre. **Whenever a road needs to be resurfaced we will seek to take advantage of the new, quieter surfacing that is available in deciding which treatment is appropriate for each location.**

4.216 We recognise that noise from commercial vehicles, especially empty lorries, can be very intrusive. Following a joint review with industry, we intend to publish a joint Guide to Best Practice advising operators and manufacturers how to minimise such noise.

Noise mitigation

4.217 It is not technically feasible to eliminate all transport noise at source and we will therefore consider ways of mitigating the effect of noise where appropriate. The European Commission has proposed a Framework Directive on environmental noise to harmonise methods of calculating noise exposure so that targets and action plans could be developed, initially by Member States but eventually at EU level. We are playing an active part in the technical groups that are taking this work forward.

4.218 We recognise the significant impact of increased traffic on roads which were not designed to carry large volumes of traffic. Excessive noise is a major source of concern to people who live near busy roads and in some cases noise levels at properties are now much higher than the threshold which would trigger noise insulation if the road had been built to current standards. **We are examining the scope for noise mitigation measures on trunk roads built before these standards were created.**

4.219 The speed at which road traffic travels is also important in the level of noise which is produced and its effects on the population. **The relationship between road traffic speed, emissions, noise and**

safety will be considered as part of our review of speed policy (see Chapter 3).

4.220 The measures taken by local authorities to address significant local air quality problems or to meet local road traffic reduction targets should also result in improvements in the general environment, including reduced noise. It is important that the effect of traffic management measures on noise is taken into account in assessing proposals.

4.221 We propose to take powers to enable airports to enforce mitigation measures, for example by taking action against non-compliant airlines, and to enable local authorities to enforce noise mitigation agreements. This will require legislation and we will consult on the details. In the interim, there is nothing to stop airports from entering into voluntary noise mitigation agreements with their local authorities.

4.222 The Planning Guidance Note on noise (PPG24) advises local authorities in England on the use of their planning powers to minimise the impact of noise. It sets out criteria for permitting noise-sensitive and noise-generating developments and advice on conditions to minimise the impact of noise.

4.223 Responses to the consultation expressed concern about the application and interpretation of this guidance, especially in relation to smaller airports. We will review the operation of the guidance and, if necessary, take action to ensure that its principles are followed by local authorities and developers.

Technology – research and development

4.224 Technological development offers many new and better opportunities for integrating transport.

Our policy is to use the most appropriate and cost-effective technology for each task and to encourage pilot trials of newer technologies or systems that show special promise.

4.225 We expect the private sector to play its full part in bringing to the market transport technology which supports integration. For our part, we will seek to help remove institutional and other non-technical barriers to the use of technology, in partnership with others.

Solutions for the 21st Century: a solar-powered car designed by Honda.

4.226 Our main aim is to identify and assess all the implications (including the safety, environmental, social and financial implications) so that everyone is clear about the likely effects of implementing a specific technology or group of technologies. Although the development and deployment of some technologies are likely to occur in the normal course of business, we will take action when necessary to promote progress. For example, we will:

- consult widely on technology development and research needs;

- support research and development of relevant promising technologies – through either wholly funded research and Seedcorn programmes or collaborative schemes such as a LINK programme;

- support trials, demonstration and validation projects and pilot implementation projects;

CHAPTER 4   Making It Happen

- oversee the dissemination of research results and promote good practice;
- facilitate public-private partnerships with clearly defined roles;
- provide incentives to promote the use of cost-effective technology through fiscal and regulatory measures, and promote discussions on the way forward;
- ensure that the aims and objectives of this White Paper are fully integrated into the new Foresight 2000 initiative.[15]

Bus information: updating electronic bus timetables in Southampton.

4.227 Research and development into technology is carried out by many organisations: the European Commission, local authorities, research councils, the science base and industry. A partnership approach will often provide the best way forward. Working with our European partners, for example, we are seeking to ensure that the EU Fifth Framework Research and Development Programme supports research relevant to integrated transport and will complement national research.

4.228 Technologies develop at different rates and can be superseded relatively quickly. We believe that the best approach is one that builds on existing knowledge and development, where each step adds value and is consistent with the longer term solution.

4.229 Intelligent Transport Systems (ITS), or transport telematics, for example, is being rapidly developed by a growing sector. It could be one of the most significant transport applications of technology to emerge in recent years. It has the potential to deliver many integrated transport objectives, including comprehensive real time travel information and guidance.

4.230 Some work on the need for telematics systems to 'inter-operate' has been completed or is in progress in the USA, Japan and Europe. While not every UK system needs to operate with every other system, many may need to do so in future. Building on the results of this work, we will assist in the development of transport telematics applications, including those that are relevant for public transport, as a priority.

---

15   The Office of Science and Technology is consulting on proposals for the next round of work under the UK Foresight Programme, which the Government intends to launch in April 1999. "Foresight – consultation on the next round of the Foresight programme", DTI, March 1998.

CHAPTER 5
A Shared Responsibility

> *"For Leon's sake and for the sake of everyone, you can 'do your bit' to make his world – our world – a cleaner, better place in which to live."*
>
> Launch of the DETR campaign
> March 1998

Partnership for action

5.1 Our *New Deal for transport*, sets the framework for change – but Government cannot achieve this alone. Business, operators, communities and individuals all have a part to play in responding to the challenge. Green transport plans produced by local authorities, businesses, community organisations, schools and hospitals will alert people to the problems and the solutions. We will help to spread information about new ways of working and living which reduce the need to travel and the impact of journeys. Partnership of various forms provides a good means of bringing different interests together.

Partnership in innovation and design

5.2 We are keen to work in partnership with vehicle manufacturers and the oil industry to see what more can be done to accelerate the pace of change in vehicle technology and to ensure that there are incentives for people and businesses to buy more fuel efficient, less polluting vehicles. **We have set up the Cleaner Vehicles Task Force** with senior industry representatives, environmentalists and other organisations to take this forward.

5.3 The Task Force aims to promote the production, purchase and use of vehicles which are more fuel efficient, less polluting and quieter and to improve the environmental performance of the existing vehicle fleet. It will be an important source of advice and information for business and the public, educating and raising awareness of the benefits of greener vehicles and considering attitudes towards marketing and advertising of cars.

5.4 The Task Force is considering a range of options, including the role of voluntary targets for the purchase of vehicles with lower emissions, the scope for improvements in the enforcement regime and the need for guidance for local authorities. The Task Force expects to publish progress reports as its work develops.

5.5 We support the Foresight Vehicle Initiative (developed under the Foresight Programme – see Chapter 4) which aims to promote the development of motor vehicle technology that is significantly more environmentally friendly and capable of meeting mass markets requirements of safety, performance, cost and desirability. There is a close link to the Cleaner Vehicles Task Force and we are supporting a Foresight Vehicle LINK programme which provides for collaborative research with industry in this area.

5.6 We will pursue innovation and development in technology in partnership with organisations in the private sector, among charities and with the research community. The sponsorship of LINK programmes is one way of achieving this end. We will also be consulting widely on how the aims and objectives in this White Paper can best be promoted through research and an understanding of likely technology futures. The development of a clear understanding of transport behaviour is essential.

CHAPTER 5   A Shared Responsibility

Partnership: to help the motorist

> **SMMT – Greener Motoring Guide**
>
> The Society of Motor Manufacturers and Traders (SMMT) has recently launched a Greener Motoring Guide, which encourages drivers to use their cars less, and use them more effectively when they do. Key messages are:
>
> - try to drive less;
> - service your vehicle regularly;
> - keep an eye on fuel consumption;
> - buy cleaner fuels;
> - your driving style has a very significant effect on the emissions your car produces.
>
> The campaign will run until October '98. Copies of the booklet will be provided in all new cars until then. They also hope to make copies available in all franchised dealerships as well as through the organisations which have endorsed the guide – Institute of Advanced Motorists, Retail Motor Industry Federation, AA, RAC, BSM and Tesco.

5.7 Buying a used car can be fraught with problems for the motorist – ensuring that the car is roadworthy and has not been stolen or 'clocked' can be difficult. The Office of Fair Trading report "Selling Second Hand Cars" (October 1997) recommended a number of measures that would reduce the risks and help the consumer to make the right choice.

5.8 The Driver and Vehicle Licensing Agency will consider ways of introducing new services to help the motorist, such as the provision of further information on second-hand vehicles. The Agency will also aim to improve the speed and efficiency of its customer services more generally through further investment in technology.

Working with transport operators

5.9 Responses to the consultation from transport operators have shown that they want to work with government to implement the new policy. We share with them the aim of increasing the use of public transport that is safe, environmentally-friendly and meets the needs of their customers. We expect operators to ensure that their services are provided in a way that supports our *New Deal for transport*.

BUS DESIGN

5.10 We need to improve the image of the bus, if we are to attract people who are used to the style and comfort of modern cars. The latest generation of buses demonstrates an increasing recognition of the importance of interior design quality and comfort and the industry is currently investing in new buses at the rate of around £270 million a year – an increase of some 80% in real terms on the level five years ago. We are keen to work with the operating and manufacturing industry to promote high quality design. We want to see a modern bus which is environmentally-friendly and designed to carry people with children and shopping in comfort. **We look to the industry to respond to this challenge with a bus designed for the 21st Century**.

Working with business

5.11 Business more generally has an important stake in the changes we are seeking. We look to business to ensure that it makes the most effective use of transport in a way which supports sustainable development. This means reducing the impact on the environment and helping to reduce congestion.

5.12 We are looking to business to be our partners in tackling unemployment and social exclusion through our Welfare to Work programme. This marks our commitment to tackle long term unemployment and includes a New Deal for young unemployed people.

> **Help with transport costs for participants on the New Deal for 18-24 year olds**
>
> One problem faced by unemployed people is the additional transport costs of looking for work. We asked transport operators whether they could help young people on the New Deal for 18-24 year olds. The response has been very encouraging with companies offering a variety of help including:
>
> - at least 50% discounted fares on some travel routes/services;
>
> - a pilot scheme allowing all New Deal participants to travel for a flat rate of 49p;
>
> - up to six free travel to interview journeys per person; and
>
> - free fares for all travel to interview journeys followed by free fares for the first month of employment.

Local partnership

5.13 We want local authorities to pursue public-private partnerships, where that provides the means of securing best value. In particular, they will want to build and extend partnerships with other authorities, locally and nationally, and with transport operators, including bus and train operators and freight companies, in the form of Quality Partnerships or other arrangements to improve the provision and operation of transport services in their areas.

5.14 In this context we welcome initiatives such as the Peak Park Transport Forum's consultation on an integrated transport strategy for the Peak District National Park, which was launched in April 1998. Another example of local partnership is the work done by Cumbria County Council, the National Park Authority, Cumbria Tourist Board and the Countryside Commission to develop a transport strategy for the Lake District. We are looking to work with local partners to see how Government can help to build on this approach.

A shared responsibility: individuals, families and communities

5.15 Responsibility for changing travel behaviour will be a shared one. Our new approach will provide more choice about when and how to travel, to support the objectives of reducing congestion and pollution that we all share. But individual choices will be critical to our success in improving the quality of life and the speed at which the benefits will be felt.

5.16 This need not mean a dramatic, overnight change in the way we travel. Making one less journey a week or occasionally leaving the car at home and walking or cycling instead, may not seem much to the individual concerned. But when multiplied over and over again, throughout whole communities, the impact will be substantial.

5.17 Individual action counts, but working in partnership is also a powerful way to generate ideas, get the most out of resources and secure a wider commitment. Local Agenda 21 actively involves the local community in identifying problems and developing solutions through public discussion and participation. Local communities can come together to work for sustainable transport through Local Agenda 21 and in a variety of practical ways.

5.18 We need to change our travel habits if we are to be a healthier and more prosperous nation.

To illustrate the changes that can be made, we consider three journey types – getting to work, travel in the course of business and the 'school run'. There are many others.

Journeys to work

5.19 Major employers can play their part by preparing green commuter plans which help employees to use alternatives to driving to work alone. This can make a major contribution to easing congestion, especially during rush hours. Smaller enterprises may wish to consider what they can do to help.

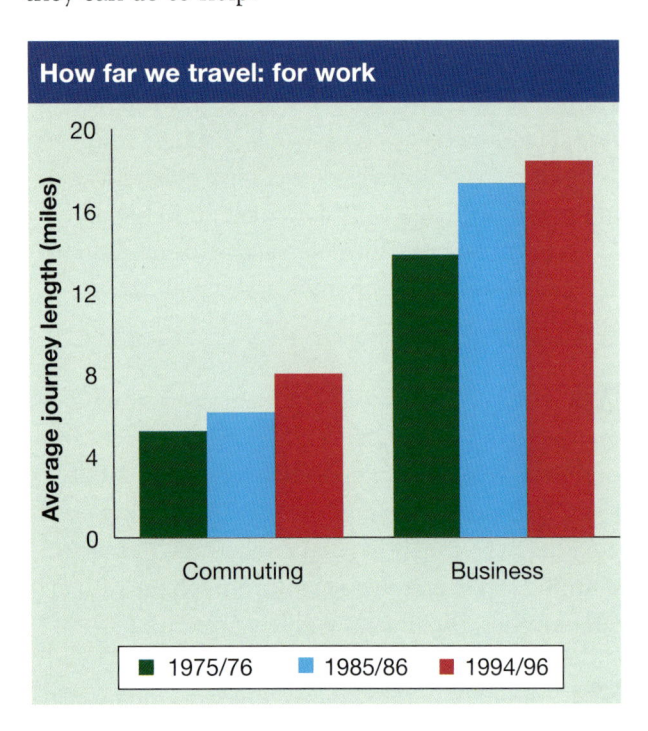

How far we travel: for work

[Bar chart showing average journey length (miles) for Commuting and Business in 1975/76, 1985/86, and 1994/96. Commuting: approx 5, 6, 8 miles. Business: approx 14, 17, 19 miles.]

5.20 Green transport plans also address business' transport use and cover travel in the course of business. **We will work with local authorities to help them secure widespread voluntary take-up of green transport plans through partnership with business and the wider community.** Part of this will involve local authorities leading by example and setting targets in their *local transport plans*.

5.21 In preparing their green transport plans, businesses can use the report[1] published last year by the Advisory Committee on Business and the Environment (ACBE). This recommended that businesses develop commuter plans and set voluntary targets for reducing single person car commuting. As a guide, **ACBE recommends that companies look to reduce by 10% the total number of people commuting to and from work, alone, by car.** We will ask the *Commission for Integrated Transport* to monitor progress on the take-up of green transport plans.

Benefits of a green commuter plan

- strengthens environmental performance and improves environmental image;

- offers substantial savings by reducing the need for workplace parking and releasing land and buildings for more productive uses;

- makes work sites less congested and more accessible for deliveries and visitors and improves relations with neighbours;

- helps staff arrive on time and with less stress by improving travel arrangements;

- attractive benefits and savings for employees enhance the recruitment package;

- promotes equal opportunities by providing travel perks throughout the organisation;

- helps staff to be healthier, fitter and more productive by encouraging exercise.

source: Transport 2000 "Changing journeys to work"[2]

---

1 Copies available from ACBE Secretariat: 0171 890 6568.

2 "Changing Journeys to Work. An employers guide to green commuter plans." Transport 2000, supported by London First.

Partnership for action

5.22 Partnership can help here, for example where local or regional public transport discounts are made available to organisations which commit to a green transport plan and appoint a staff travel co-ordinator to work on implementing and promoting it.

Green commuting.

5.23 We will take the lead by introducing green transport plans in all Government Departments and their agencies. These will cover commuting, travel in the course of work, fleet management and influencing suppliers' travel behaviour and should reflect the advice provided in our "Guide to Green Transport Plans". **We have set a target that all headquarters buildings and main buildings occupied by Executive Agencies and Government Offices for the Regions should have green transport plans by March 1999 and all other key buildings by March 2000.**

Green commuting in practice

Boots green commuting plan:

- aims to reduce car commuting to the company's Beeston site by 10% by 2000 and a further 10% by 2005;

MOD at Abbey Wood Bristol:

- new Procurement Executive HQ includes a new railway station, built in partnership with local authorities;
- a wide range of bus services – 50 buses an hour during peak periods and regular buses to the mainline station to London;
- cycle paths and facilities for cyclists – over 300 staff regularly cycle to work;

Birmingham's 'Company Travelwise':

- supported by the Passenger Transport Executive (CENTRO), the City Council and transport operators;
- most of the major local employers say they will join, benefits include:
    - discounts on travel passes, bus travel and taxi hire;
    - help in negotiating improvements to local services to meet staff needs.

5.24 We are particularly keen that hospitals are seen to be taking the lead in changing travel habits. By the very nature of their work, hospitals should be sending the right messages to their communities on acting responsibly on health issues. We would like to see all hospitals producing green transport plans.

> **Greener business travel: Parkside NHS Trust, London**
>
> The Trust has agreed a green business travel policy for all new staff. Features include:
>
> - cycling on business attracts reimbursement at 15p per mile, and walking 7p per mile;
>
> - passenger rate increased from 2p to 10p per mile to encourage car sharing;
>
> - flat mileage rate for cars regardless of engine size, higher mileage rate for Liquid Petroleum Gas vehicles;
>
> - provision of lease cars only for those who drive more than 3,000 miles a year on business, need the security of a car for regular work out of hours or regularly carry heavy equipment;
>
> - examples of good practice like this will be included in the **Healthy Hospital Toolkit** – a Transport 2000 guide to reducing car trips to NHS facilities.

5.25 Industry and business have a substantial impact on travel patterns in the surrounding area but many other organisations also generate large numbers of journeys, including hospitals, and institutions of higher and further education. Travel to leisure facilities and visitor attractions is another important component of overall travel. Reducing car use for access improves relations with neighbours and may be a condition of expansion.

> **How far we travel: for shopping and leisure**[1]
>
> [Bar chart showing average journey length (miles) for Shopping and Leisure across 1975/76, 1985/86, 1994/96]
>
> [1] Journeys within Great Britain

Teleworking

5.26 Businesses may wish to consider the extent to which teleworking can reduce travel by allowing employees to work at home or at a 'satellite' work centre closer to home. This is relevant to green transport plans, particularly when teleworking can substitute for high-mileage driving patterns. Where staff spend a lot of time driving to clients, or places of inspection, in the course of their work, teleworking also has potential.

5.27 We support the use of teleworking for reducing travel, but it can give rise to social and possibly regulatory issues that should be taken into account. The benefits for the environment will also be lower if teleworking is offset by increased car travel from home. It also risks encouraging movement out of towns into the countryside, prompting less sustainable travel patterns overall.

5.28 We will therefore focus efforts on communicating best practice and encouraging local authorities to support teleworking (including through the sensitive application of their

Partnership for action

development control responsibilities), where this will substitute for high mileage car travel.

Teleworking – reducing reliance on the car

- Hertfordshire County Council – developed 'oases', localised workstations, for trading standards officers to cut out journeys to headquarters. Saved 5-8% in travel costs and 7% in car mileage;

- RM Consulting – introduced a pilot scheme in 1995. They now have 145 'location independent workers' hot-desking and much work being done from home. They plan to have 300 by the year 2000. They estimate that over 10% of these workers travel half the mileage they did in 1995. Total mileage reduced by approximately 0.5 million kms;

- ADAS Consulting – introduced IT-based working practices and reduced office sites from 90 to 25. More than 500 staff now work permanently from home and more than 1,200 use internet e-mail systems. Each home-based consultant is estimated to have reduced car use in the course of their work by 2,000 miles a year.

School journeys

5.29 We know that the issue of the 'school run' concerns many. The concern goes deeper than a wish to reduce congestion by discouraging parents from taking their children to school by car, although the benefits for the morning rush hour would be considerable. Not walking or cycling to school means that children get much less exercise and builds in car dependency at an early stage in a child's development. These children will find it harder as adults to use cars responsibly and will have fewer opportunities to develop the road sense they need as pedestrians or cyclists.

5.30 We understand parents' concerns about the safety of their children and that for many using a car has become the only way to manage a tight schedule. **Our policies will help reduce the need for children to be driven to school by encouraging safer routes for walking and cycling, giving greater priority to public transport and, through our planning policies, improving opportunities to get to work, shops and other facilities without having to use the car.**

Safer routes to school.

CHAPTER 5 A Shared Responsibility

5.31 We will continue to take account of transport issues when shaping Government policies which relate to children's journeys to school, for example by developing healthy schools initiatives that include safer routes to schools.

Safer routes to school in St Albans

Two schools in St Albans working on a safe routes to school project – Sandringham Secondary School and Wheatfields Junior – have adjacent sites and similar travel problems. They want to improve safety and encourage parents and children to use alternatives to the car for school trips. Results so far include:

- development of a school transport plan, including modal shift targets;
- a new bus service, used by 35-40 pupils a day, including 30 who used to come by car;
- 110 junior pupils trained for on and off road cycling (25% of final junior year now cycle to school);
- two new puffin crossings in place;
- involvement from 50 local volunteers, including escort schemes for children going to school by bus or on foot.

5.32 Several local authorities are already doing valuable work aimed at reducing car use for journeys to school. We are providing support in a variety of ways – for example by:

- **helping to fund the Sustrans 'Safe Routes to School' demonstration projects** in Leeds, York, Colchester and Hampshire. These encourage children to walk or cycle to school;
- **funding specific projects** in West Sussex, Manchester, Birmingham, Warwickshire and in London.

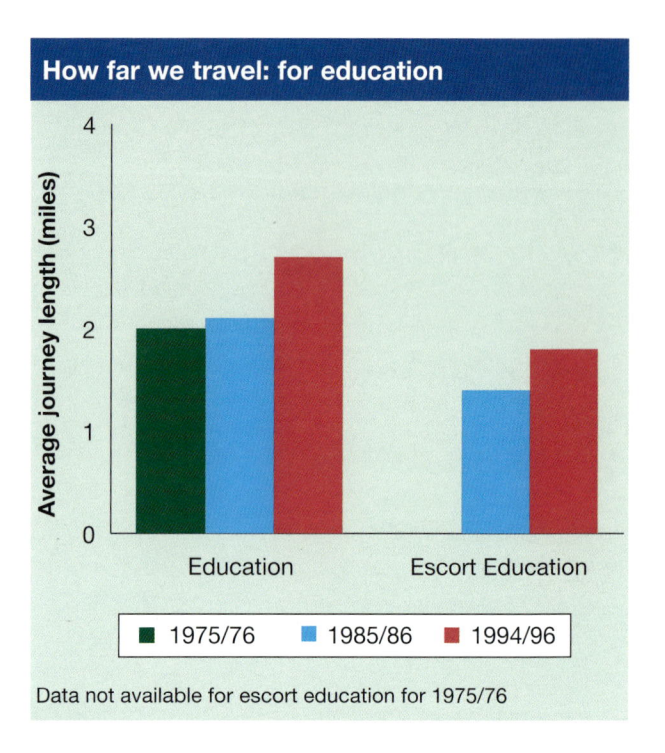

How far we travel: for education

Data not available for escort education for 1975/76

5.33 School crossing patrol officers have an important part to play in helping road safety around schools. **We will bring forward legislation to strengthen their powers so that they can help children below school age and adults to cross the road.** The legislation will also extend the hours during which local authorities can provide school crossing patrols, so that they can tailor provision to local needs.

5.34 We will build on the best of current practice and help local authorities, schools, parents and teachers develop a comprehensive approach that reflects local needs and views. Measures that could be considered include escort schemes, before and after school clubs, adjustments to the school day, improvements to local transport services, traffic management and school facilities for cycling. **We will take further initiatives to encourage more children to get to school other than by car. These will include:**

- setting up a School Travel Advisory Group with Government Departments, local authorities and others to lead the dissemination of best practice and to contribute to the development of policy;

- encouraging local authorities to include measures and targets to reduce car travel to school in *local transport plans*;
- distributing guidance on best practice for promoting alternatives to the car and on developing green transport plans;
- encouraging schools and local authorities to take account of the transport implications of their educational policies;
- encouraging communities to reduce car use without compromising safety, in ways which actively involve children, school governors, parents and local business;
- securing private sector support for school transport initiatives, building on the recent initiatives to fund computers;
- as announced in the Healthy Schools initiative, including measures to encourage safe alternatives to the car for travel to school in the criteria for the 'Investors in Health – Healthy Schools Award';
- covering school journeys in broader national awareness campaigns.

Building communities

5.35 Conventional public transport cannot always meet the diverse accessibility needs of all in our communities, particularly the needs of disabled people and those who live in remote rural areas.

5.36 Voluntary action is a strength of local communities everywhere. In London, for example, it has given rise to an extensive network of transport services run on a voluntary basis for disabled people. **We are conducting a review of voluntary and community transport activity**. There are already relaxations of the normal rules for bus operator licensing to help non-profit making bodies, especially those who provide 'community' bus services including mini-buses. The review will provide a better understanding of the role played by the voluntary sector and allow us to consider whether policies at local or national level should be changed to enable the voluntary sector to operate more effectively.

Community transport charter

The Community Transport Association and the Transport and General Workers' Union have launched a minimum standards charter, aimed at all those who fund and operate community transport services, both paid and voluntary. Key points are:

- regular training for drivers and assistants;
- assessment of health and safety of workers, including driver stress and fatigue;
- training in safety and help for passengers, especially for children, disabled people and elderly people;
- attention to vehicle safety and maintenance.

5.37 In the countryside voluntary action has supported flexible and innovative approaches to meeting the increasingly diverse needs of rural communities. In preparing *local transport plans*, local authorities will need actively to involve their local communities, to ensure the right balance of priorities is struck.

5.38 Parish councils in England both through their local knowledge and commitment, and through their new powers to fund transport projects by raising money through a precept on council tax, could be valuable partners in improving local accessibility in rural areas. We would like to see them take an increasing role in community transport – using their powers to survey transport needs and to fund community bus services, car sharing schemes and concessionary fares for taxis.

Car clubs

- owning a car is expensive but individual journeys can seem relatively cheap. Once a car is acquired, it acts as a disincentive to using public transport;

- the 'city car club' is one solution which has proved very successful in Europe. A pilot starts this summer in Edinburgh: ownership and use of cars is shared – to provide a car when it is really needed but avoid unnecessary use. This is different from conventional car hire in that the cars are kept locally and can be used at short notice and for short periods of time;

- experience from Germany is that members of clubs who were previously car owners reduce their mileage by half. City car clubs can also help to reduce pressure for parking spaces.

5.39 Many of the more innovative proposals in England have been supported by the Rural Transport Development Fund, which is administered by the Rural Development Commission (RDC). We have increased the level of resources going into the fund but the RDC still has to turn away worthwhile projects.

5.40 We intend to build on the success of the Rural Transport Development Fund in England by **creating a new *Rural Transport Partnership* scheme** to run alongside. This will help to get extra resources into rural transport where it counts.

5.41 The new scheme will enable parish councils and voluntary groups to work in partnership with local authorities. The aim is to support schemes which reduce rural isolation and social exclusion through enhanced access to jobs and services. These will be based on local needs and the local community should participate in their development.

5.42 We plan to support the new initiative with **£4.2 million a year, additional to the resources for transport already going into the countryside in England**. Successful projects will be those that galvanise local initiative and offer the prospect of long term enhancements to the quality of rural transport. A key theme will be better co-ordination of existing voluntary, local authority and commercial services.

Improving rural transport

The Snowdon Sherpa:

- buses can help to reduce traffic congestion in rural 'honeypot' locations. The Snowdon Sherpa in the Snowdonia National Park is one of Britain's longest established national park schemes;

- major factors in its success are close co-operation between Gwynedd Council, the National Park and local operators and good publicity for a network of services;

Moorlands community minibus:

- a self-help project providing a scheduled door-to-door service to help elderly residents reach the services they need or to shop in the nearby towns such as Leek or Ashbourne;

Cheshire Rural Rider:

- provides accessible bus services around Macclesfield, helping rural residents to get to local day centres or nearby towns. It has pioneered the successful integration of local authority social service and public transport responsibilities.

Supporting your local railway

5.43 Rural rail services provide an alternative to the car and for some journeys one that is not easily substituted by bus. We consider that many of them

Partnership for action

are not delivering their full potential, for a variety of reasons. Our proposals for better information for public transport users, for better integration between different forms of transport and for easier ticketing can help encourage more people to use trains in the countryside (see Chapter 3).

5.44 We also wish to see more local initiatives, particularly community-rail partnerships, where local businesses get involved in packages to promote leisure and tourism by rail and more regular use by local people. The support of local authorities can also be critical in developing local stations as hubs of economic activity and social interaction.

The Esk Valley Partnership

Focuses on the Middlesbrough-Whitby line and is funded in part by the European Union, local authorities, Regional Railways North East and the Rural Development Commission. Achievements include:

- improved publicity and signposting of stations;
- greater co-operation between rail and bus operators;
- a residents' railcard giving discounted travel;
- on-train events; and
- school and community projects, with 'adoption' of stations by community groups.

5.45 The Countryside Commission is running consultancy research to test new ideas for rural interchange sites called 'Staging Points' aimed at bringing transport back to the community. These sites will be places in the countryside and urban fringe with some car parking, such as rail stations, village halls, community centres, leisure and visitor attractions. These places would provide the focus for a greater variety of opportunities to interchange and link to other transport networks might be provided, including the provision of bicycles and better bus and taxi services. Added value will come from community-centred design of transport and from local enterprise services such as home shopping distribution and cycle hire or repair.

Raising awareness and informing choice

5.46 We will support individual and community choice by improving information and awareness of the impacts of different ways of travelling. We will promote a climate where the effects of those choices, on the individual, on their environment and on others, are better understood.

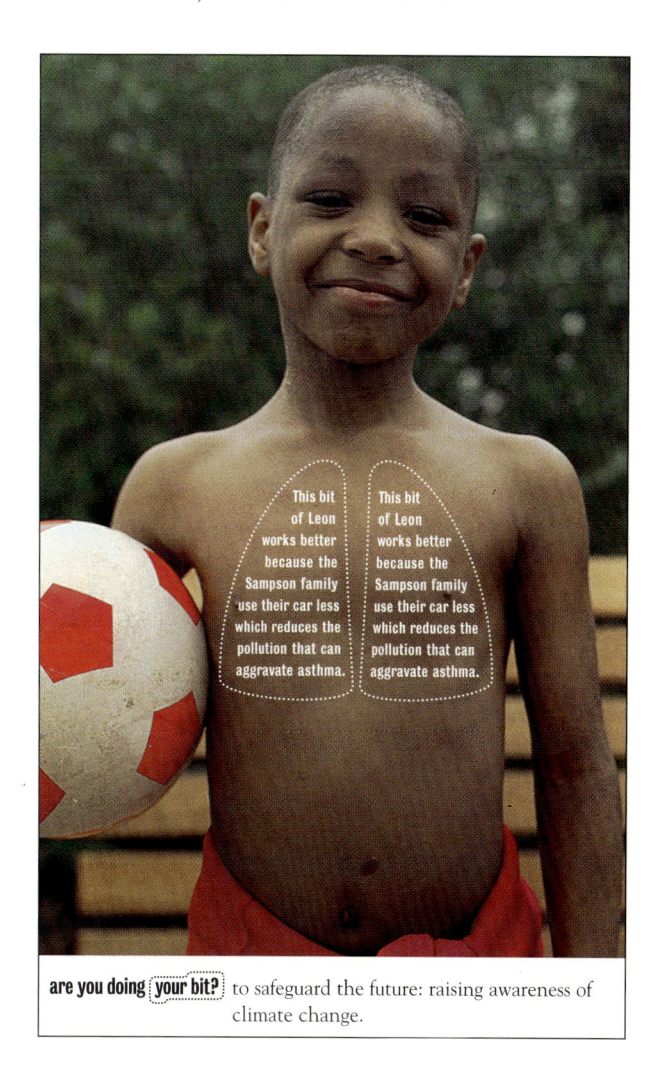

are you doing your bit? to safeguard the future: raising awareness of climate change.

CHAPTER 5   A Shared Responsibility

5.47 Many local authorities have travel 'awareness' campaigns, often with quite limited resources. Their main aim is to increase recognition among local people that there is a need to reduce the environmental impacts of car use. The campaign provides a climate in which specific measures aimed at achieving this, whether voluntary or through regulation or charging, can be accepted. We strongly welcome these initiatives.

5.48 Most local authority campaigns are branded under the logo 'TravelWise'. Some local authorities have given particular emphasis to encouraging young people to be environmentally aware; for example, by making the most of opportunities in the school curriculum to consider environmental issues.

Travel awareness campaigns

- 'TravelWise' – a local authority travel awareness initiative, started by Hertfordshire. Activities include local advertising (local radio, leaflets etc), working through local groups, schools packs, plus public transport travel information. A National Travelwise Association was launched in March;

- 'Don't Choke Britain' – a national campaign held in June each year: encourages car commuters to try something different at least one day a week in June – to use public transport, cycle, walk, share a car or travel at outside the rush hour. Acts as an umbrella for other campaigns including National Bike Week, Walk to School '98, Green Transport Week and the Car Free Day;

- Association for Commuter Transport – develops and promotes sustainable travel initiatives, provides employers with advice, education and training opportunities and a forum for exchanging ideas and best practice.

5.49 The results of two EU-funded projects, INPHORMM[3] (for which the University of Westminster is project co-ordinator) and CAMPARIE[4] (in which Transport and Travel Research (UK) is a partner), will develop our understanding of the effectiveness of local awareness campaigns, and help to get the most out of them. Local authorities' action is given additional weight and impetus through Government funded campaigns at the national level, such as the 'Going for Green' campaign and the 'Are You Doing Your Bit?' campaign launched in March 1998.

Are You Doing Your Bit? by

- leaving your car at home for at least some journeys;

- walking and cycling more and making more use of buses and trains;

- getting a garage to tune your car properly and making sure tyre pressures are correct.

Going for Green

- a Government-supported campaign to inform people of lifestyle changes that can make a difference. It has produced a five point Green Code of steps that everyone can take, including one on transport;

- this year the Green Code is being promoted by 'theme months': 'Travel Sensibly' is in June, when Going for Green plays a key role in the 'Don't Choke Britain' campaign.

5.50 We will continue to fund publicity campaigns at the national level to raise awareness of how small changes in personal behaviour and lifestyle can make for a better environment. We will look for fresh ways to highlight the link between individual consumption and the threat to

---

3   Information and Publicity Helping the Objective of Reducing Motorised Mobility.

4   Campaigns for Awareness using Media and Publicity to Assess Responses of Individuals in Europe.

global climate as well as to the quality of the local environment. Campaigns will aim to show that changes in travel behaviour which are good for the environment do not involve lifestyle sacrifices and will stress the personal benefits, including those for health, of using cars less.

A New Direction

5.51 This White Paper signals a new direction for transport in which everyone must play a part if we are to succeed. Many of the changes can start immediately and, as we have illustrated in the examples of good practice, much can be achieved without the need for legislation. Over the longer term, new sources of funding will provide a further impetus to these reforms.

5.52 We cannot expect people, business and communities to make changes in their own use of transport if they do not understand what difference it makes. **We are committed to the reforms set out in this White Paper and we will publish information on how successful the new approach is, measured against our targets and objectives, over the coming years.**

5.53 We should not wait another 20 years before reviewing transport policy. The *Commission for Integrated Transport* will play an important role here – in monitoring progress, in bringing together different interests and in advising on further action in the light of changing circumstances – some of which we cannot now foresee. Through this process, we will be able to update and review the strategy and measures set out in the White Paper in the light of developments, to secure the changes that we all want to see. This will be a vital part of our *New Deal for transport*.

CHAPTER 5   A Shared Responsibility

The *New Deal for Transport* will make a big difference to all our lives.

ANNEXES

ANNEX A   Future publications

ANNEX B   Consultation on integrated transport policy

ANNEX C   Royal Commission on Environmental Pollution

ANNEX D   'Transport: The Way Forward'

ANNEX E   Core trunk road network map

ANNEX F   Rail network pinch-points

A New Deal for transport: better for everyone

ANNEX A

Future publications

Integrated Transport Policy – associated publications

The following papers will set out in more detail the proposals in the White Paper and are expected to be published shortly.

- Trunk roads policy: outcome of the reviews for England and Wales
- Railways policy: a response to the third report of the Environment, Transport and Regional Affairs Committee on the proposed Strategic Rail Authority and railway regulation
- Bus policy
- Charging policy: a consultation paper on implementing road user charging and workplace parking charges
- Shipping policy: a response to the recommendations of the Working Group on Shipping
- Freight policy: a paper on sustainable distribution
- Road safety policy: strategy and targets for beyond 2000
- Guidance on local transport plans
- A report on inland waterways

OTHER RELEVANT DOCUMENTS TO BE PUBLISHED INCLUDE:

- A response to the Royal Commission on Environmental Pollution's report, "Transport and the Environment – Developments since 1994"
- A consultation paper on Climate Change
- A report on the review of the National Air Quality Strategy
- A revised strategy for Sustainable Development
- Consultation on draft or updated guidance for:
  - producing better development plans, describing how they integrate with local transport plans (revised PPG12)
  - the new approach to regional planning (new PPG11)
  - land use and transport (revised PPG13)
- An action plan to encourage walking
- Progress reports from the Cleaner Vehicles Task Force

ANNEX B

Consultation on integrated transport policy[1]

Between 21 August and 14 November 1997, the Government carried out a major consultation on the integrated transport policy throughout the UK. Over 7,300 responses were received in a written consultation and a number of consultation meetings and seminars were held throughout the UK. Analysis of the written responses and of comments made at the meetings has shown that there is a clear consensus on a number of issues:

- there was overwhelming agreement that it is time for a change in the direction of transport policy. People want more choice;

- many people wanted more and better facilities for pedestrians and cyclists and for walking and cycling to be treated as modes of transport in their own right;

- a significant number of responses emphasised the health benefits of a reduction in car use. Many wanted to see people walking and cycling more, for road safety to be improved and for people who do not have a car to have better access by public transport to jobs and services. Others were concerned about transport's impact on local air quality and on noise levels;

- there was a high level of support for better public transport. A general plea for more investment in transport was also made. Many respondents thought that an increase in investment could come from new, dedicated sources of income like congestion charging and a tax on private non-residential parking spaces. Businesses, local authorities and other key organisations in particular wanted money raised through these measures to be spent on transport;

- there was a general recognition that rural areas have particular transport problems which needed to be considered carefully;

- most people accepted that we cannot tackle congestion and pollution by simply building more roads. Many regions did, however, want 'bottlenecks' tackled, local road safety to be improved, some bypasses to be built and better traffic management;

- many wanted to see less freight moved by road and more to be moved by rail or coastal shipping. They also wanted existing regulations on, for example, lorry weights, speed and drivers' hours to be enforced more effectively. In addition, there was general agreement that better enforcement of vehicle standards, traffic speeds and parking restrictions would have an immediate impact and would help to improve local air quality, road safety and congestion;

- it was stressed that transport must be integrated with other policies, particularly with land use planning which has an important role to play in reducing people's need to travel; and

- there was general agreement by local authorities and transport professionals that carefully aimed education and awareness campaigns could be effective ways of changing people's attitudes about cars and could make using public transport or cycling or walking more acceptable alternatives.

---

[1] A full report is available: "The Government's Consultation on Developing an Integrated Transport Policy: A Report", published by the DETR, 1998.

ANNEX C

Royal Commission on Environmental Pollution

"Transport and the Environment – Developments Since 1994"[2]

The RCEP's twentieth report on transport and the environment set out its views on the future direction of transport policy. The main conclusions were that:

- forecast traffic growth is economically, environmentally and socially unacceptable;

- fuel price increases and improvements in vehicle technology so far planned will not in themselves bring about the requisite improvements in air quality or reductions in emissions of greenhouse gases;

- there is a need for rapid innovation in vehicle technology; better integration of public transport systems; better integration of transport and land use planning; better traffic management policies; and policies to encourage modal shift.

The Commission's detailed conclusions included:

FUEL EFFICIENCY AND EMISSIONS

The Government should make more use of **economic instruments** to encourage use of fuels which are less damaging to the environment, and to reduce fuel consumption. It should set increasingly challenging targets for reducing transport carbon dioxide emissions and, if necessary, support EC legislation to limit carbon dioxide emissions from cars.

INTEGRATION OF PUBLIC TRANSPORT

An **integrated public transport system** – primarily focused at the local level, though with appropriate recognition of the regional dimension – would offer many advantages. It requires reliability and quality, with availability of connecting services and physical provision for them; priority for public transport within the road network; good information about timetables and fares; through ticketing; and provision for people with reduced mobility.

Buses should have a central role in an environmentally sustainable transport system, and local authorities need stronger powers to ensure they provide a quality service.

Charging for use of roads within specified areas, and control of private non-residential parking would increase the efficiency with which the road network is used, and reflect more clearly the environmental and social costs involved. The introduction of charging should be a local decision and any revenue raised should be spent locally to finance public transport and other infrastructure improvements.

---

2   The Royal Commission on Environmental Pollution's Twentieth report, "Transport and the Environment – Developments since 1994" was presented to Parliament by Command of Her Majesty in September 1997. ISBN 0-10-137522-0.

ENCOURAGING USE OF ALTERNATIVE MODES

Rail regulation should facilitate strategic development of rail, and ensure that it contributes fully to an integrated public transport system

Traffic calming, the provision of quality networks, and better facilities for cyclists will be needed to encourage **cycling**.

TRANSPORT AND LAND USE PLANNING

Transport plans should be integrated with land use plans to reduce the need to travel, distances travelled, and dependence on lorries. Plans should be long term, and co-ordinated at the regional level.

FREIGHT

Initiatives to reduce freight intensity, and to transfer **freight** from road to rail should be supported.

The trend to heavier lorries appeared inconsistent with this objective, so they should only be permitted on suitable roads.

TRUNK ROADS

The emphasis should be on making maximum use of capacity of existing **road** network, removing bottlenecks through minor construction work and improving traffic management.

Any introduction of **motorway tolls** should be accompanied by measures to avoid or minimise diversion. Lorries using motorways might pay a vignette 'fee'.

INVESTMENT

An enhanced programme of **investment** over a 10 year period is needed in order to create an environmentally sustainable transport system.

OTHER

Targets for **traffic reduction** must have a clear and specific justification and must set out the preferred and most effective method of achieving those objectives.

There should be a concerted campaign to change **public attitudes** to cars.

Specific policies are needed to deal with the transport problems of **rural areas**.

The proposed **Greater London Authority** will be well placed to improve London's transport system. The responsibilities of passenger transport executives in other connurbations should be extended to cover all aspects of integrated transport, including regulation of private transport.

ANNEX D
"Transport: The Way Forward"

*An extract from the **previous Government's Green Paper "Transport: The Way Forward"** taken from Section 1 "Why transport policy has been reviewed: objectives and pressures"*

INTRODUCTION

This paper fulfils the Government's commitment to draw together the threads of the national debate on transport policy. Besides analysing the conclusions of the debate, the document sets out the Government's view of the future direction of policy – including proposals for some significant new measures aimed at developing policy to meet changing expectations. The paper concentrates on domestic, surface transport in England. The Secretary of State for Northern Ireland has produced a statement on transport priorities there; the Secretaries of State for Scotland and Wales intend to publish transport policy statements later.

KEY THEMES

The debate has shown there is growing public awareness of the impact of traffic growth, and divergence of views about how to promote sustainable development and the competitiveness of UK industry, while preserving freedom of choice. These concepts are discussed further in chapters 3–5. There are still strong demands for improved access and efficient transport, but individuals see a clear need for further action to reduce the environmental impacts of transport; and business is very concerned about the prospect of increased congestion on industrial costs. There has been a significant increase in the amount of concern on both these issues in recent years, although concern is somewhat lower in places where congestion and pollution are less severe.

Over the last few years the Government has adopted a wide range of measures to address these issues:

- since 1979 more than £24 billion has been spent in tackling congestion through investment on the national motorway and trunk road network;

- there has been very substantial new investment in public transport, including electrification of the East Coast Main Line, light rail schemes and the Channel Tunnel;

- there has been a substantial programme of liberalisation and privatisation for transport leading to wider and better services for transport users;

- new sources of funds have been tapped through the private finance initiative (PFI) for projects such as the Dartford Bridge and the Heathrow Express;
- the most harmful transport emissions in urban areas are set to fall to less than half their 1990 levels by 2005 as a result of measures already introduced to tighten regulations for new vehicles and improve enforcement for older vehicles – new measures will reduce pollution still further;
- we are on target for returning CO_2 emissions to 1990 levels by the year 2000;
- new arrangements have been introduced for local transport funding, encouraging local authorities to take a more strategic view of local transport needs. New planning guidance has been introduced, stressing the aim of reducing the need to travel.

But more needs to be done, building on these achievements. Taking further action may require hard choices be made. Policies which impact on the use of cars and lorries would need to be justified by the benefits achieved. Changes will take some time to have full effect.

Key themes which have emerged in the national debate, and which the Government proposes to pursue, including the following:

– Better planning of transport infrastructure

Concerns about both the direct and the indirect impacts of new roads have been clearly expressed, and continue to be a source of conflict. The Government invites comments on a **proposal for integrating more closely the regional land use planning system with the planning of trunk roads**. The proposal could help assessment of the needs for new infrastructure at a regional level. These arrangements will help ensure that cost-effective public transport alternatives to road improvements and traffic management options are considered at an early stage as an integral part of the decision-making process, without holding up necessary improvements. They will help to involve local interests more.

– Making more efficient use of existing infrastructure

Business and environment interests are united in wishing to see better use of existing infrastructure, both to help business competitiveness and to reduce the need for further road improvements. The Government has the needs of the business community firmly in mind. Specific measures to be pursued by Government include development of techniques for managing traffic flows on motorways; development of route strategies by the Highways Agency; improved information to road users; and carefully-targeted investment in the national roads network. Privatisation of the railways will help make better use of rail infrastructure. The Government aims to exploit new technologies where these can help promote greater efficiency. Increasing the efficiency of freight transport – on which industry depends – will need special attention.

– Reducing dependence on the car, especially in towns; empowering local decision making

60% of journeys by car drivers are less than 5 miles. Concerns about congestion, air pollution and the other impacts of car use have focused attention on the need to promote alternative transport modes: but pressures are clearly much greater in some parts of the country than in others. The need for local solutions means that local authorities must play a leading role. They already have a range of powers. There has been general support for Government planning guidance and for Government policies for funding local transport: the Government will therefore continue to support local authorities through these instruments. The Government believes there is a case for giving local authorities some additional powers to manage traffic.

– Switching emphasis in spending from roads to public transport

The debate showed a strong preference for improved public transport over expanded road capacity. The Government believes there needs to be a shift in priorities to reflect this. All transport improvements, including public transport, have to be constrained by what is effective and affordable, though privatisation and the PFI open new possibilities. Following privatisation, investment in rail will no longer be wholly dependent on public spending. Promoting bus use is likely to have a major role: the Government will implement the proposals of the Bus Working Group and help local authorities deliver bus priority measures. New strategies for promoting walking and cycling are being developed.

– Reducing the impacts of road freight

Business depends increasingly on road freight and the Government is committed to a competitive business sector, but there are growing concerns about the effects of congestion and environmental impacts. Many calls were made for increasing the proportion of freight moved by rail. The Government recognises the benefits that this can bring – both through easing congestion and through reducing environmental impacts. The Government therefore intends to put increased emphasis on encouraging alternatives to road freight, including both rail and water-borne transport. The sale of the Trainload Freight Companies is a major change in this area and will allow those businesses to pursue innovative, customer-orientated strategies to encourage freight carriage by rail. However, this will not in itself substantially reduce road freight levels unless there are also substantial changes in road delivery patterns. The Government believes that much greater effort needs to be made to reduce transport intensity. The CBI have also acknowledged this in their recently published report "Moving Forward". The Government will therefore be discussing with industry and the CBI the best means of achieving this, including greater dissemination of best practice.

In pursuing these key themes, the Government intends to develop principles which have been applied with considerable success previously in the transport sector and elsewhere. In particular, the Government will seek to improve the efficiency of markets in transport and expand their role, and to expand the role of the private sector. Within this framework the Government also seeks to ensure that the taxpayer gets value for money from those services best provided or funded by the public sector. There may be scope for developing the role of prices: especially to relate more closely to the wider costs of transport. And it will be important to ensure transport decisions are taken at the right level.

DEBATE: PUBLIC ATTITUDES

The debate was warmly welcomed. It built on the Royal Commission report: other significant reports which contributed to the debate included two from the CBI, contributions from many local authorities and their associations, the AA, the RAC, the Freight Transport Association, the Council for the Protection of Rural England, Transport 2000 and many professional institutions. Many seminars were held around the country in an attempt to reach consensus.

The Government also looked at other evidence on public attitudes. Particularly helpful were the British Social Attitudes surveys; Lex reports; RAC and AA work; and a new study carried out by Westminster University. (For references see chapter 2.) Annex 2 of the main paper considers these reports more fully.

The combined message from these surveys and from the national debate is the need to take far greater account of environmental impacts of transport, and the need for tougher measures to solve the problems of congestion and pollution without relying on more or bigger roads. At the same time we must not damage competitiveness.

ANNEX E
Trunk road network

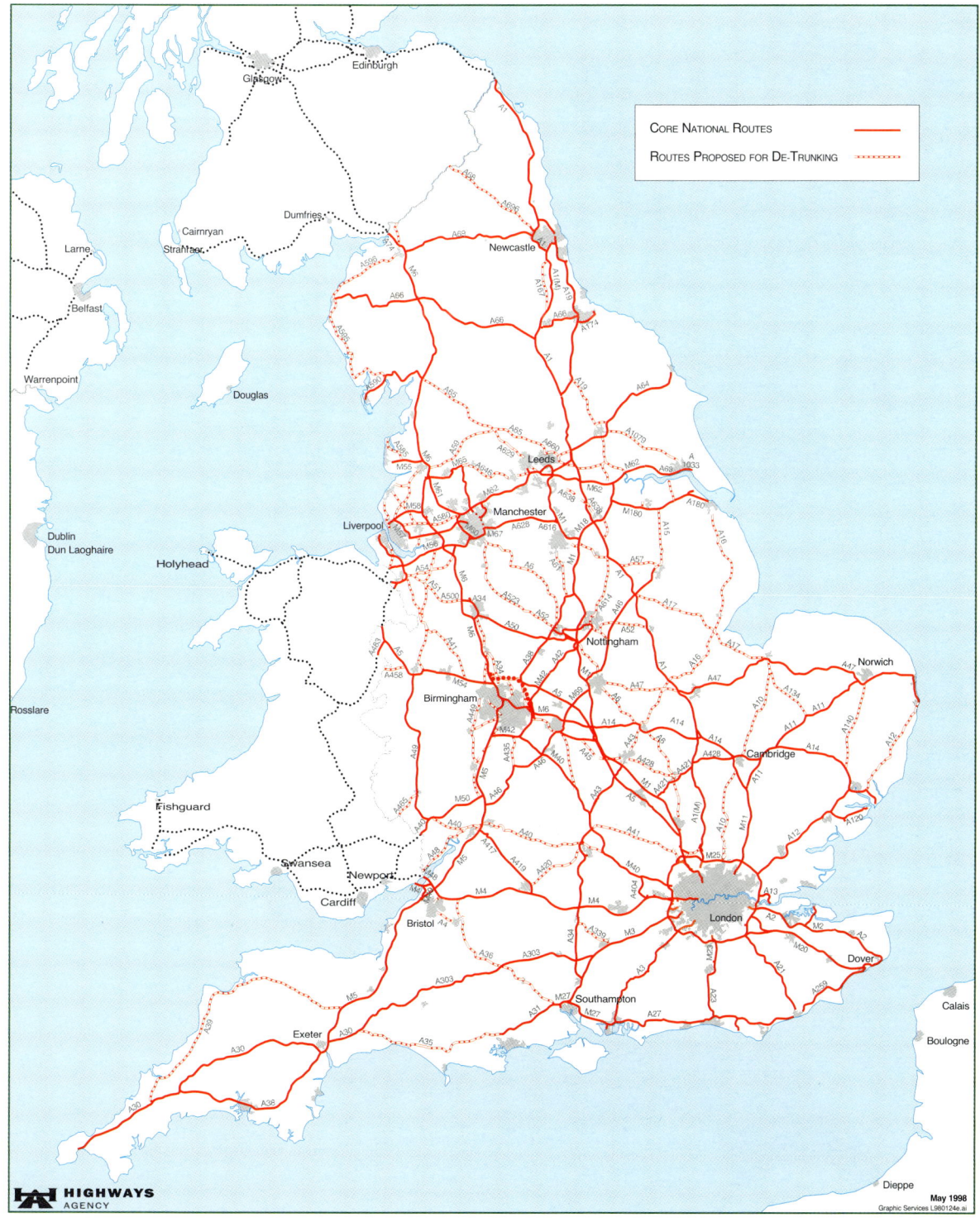

The core trunk road network in England

ANNEX F
Rail network pinch-points

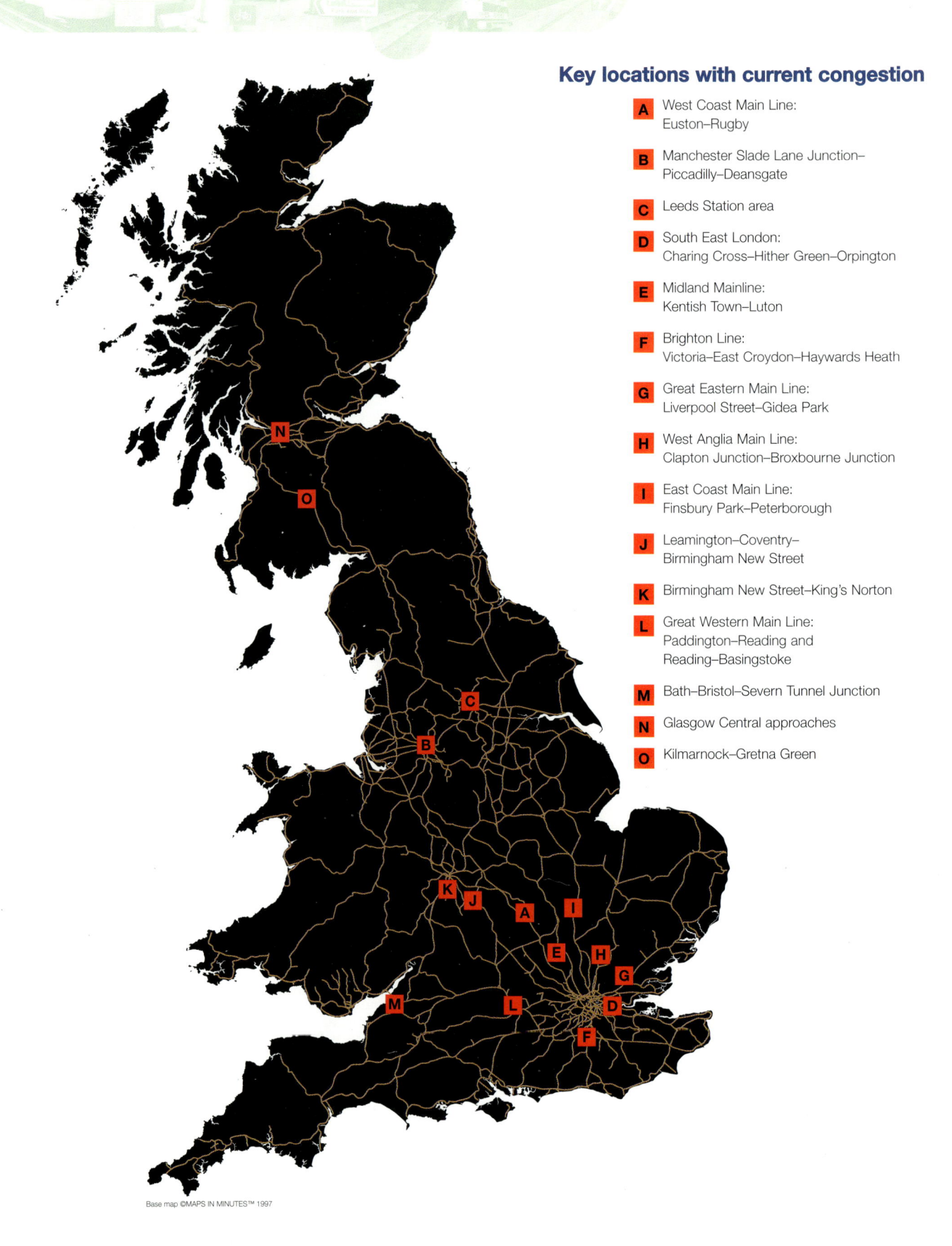

Key locations with current congestion

- **A** West Coast Main Line: Euston–Rugby
- **B** Manchester Slade Lane Junction–Piccadilly–Deansgate
- **C** Leeds Station area
- **D** South East London: Charing Cross–Hither Green–Orpington
- **E** Midland Mainline: Kentish Town–Luton
- **F** Brighton Line: Victoria–East Croydon–Haywards Heath
- **G** Great Eastern Main Line: Liverpool Street–Gidea Park
- **H** West Anglia Main Line: Clapton Junction–Broxbourne Junction
- **I** East Coast Main Line: Finsbury Park–Peterborough
- **J** Leamington–Coventry–Birmingham New Street
- **K** Birmingham New Street–King's Norton
- **L** Great Western Main Line: Paddington–Reading and Reading–Basingstoke
- **M** Bath–Bristol–Severn Tunnel Junction
- **N** Glasgow Central approaches
- **O** Kilmarnock–Gretna Green

Base map ©MAPS IN MINUTES™ 1997

INDEX

air quality
    see also pollution
    improved 35, 124–5
    in-car 11, 23
    National Air Quality Strategy 31, 124–5
airports
    access to 46, 58
    air freight 74
    air safety 88–9
    Airport Transport Forum 78–9
    BAA integration plans 80
    Civil Aviation Authority 88–9, 101
    disabled people 58
    integrated transport 76–80
    interchange policy 78
    local connections 78–9
    National Air Traffic Services 89, 101–2
    national connections 79–80
    new policy 76–8
    noise 23, 25, 136
    regional 77–8
    regulation 101
appraisal 132
Are You Doing Your Bit? campaign 150
aviation
    see also airports
    CO_2 emissions 25
    environmental impact 125
    investment 101–2
    regulation 101
axle weights
    buses 56
    lorries 71–2

British Transport Police 131
Buchanan Report 11
business 140–1
buses
    see also road traffic
    axle weights 56

Bus Appeal Body 42
deregulation 28
design 140
disabled people 56–8
drivers' hours 86
easier access 56–8
enforcement 129
environmentally friendly 120, 121
fares and ticketing 43, 48–9
Fuel Duty Rebate 113, 121
funding 112
interchange facilities 49–51
local transport plans 42
London bus lane enforcement 130
London bus priority network 106
national public transport information systems 52
passenger information 51–3
personal security 55
priority 42, 59
Quality Contracts 41
Quality Partnerships 15, 40–1, 112
road safety 85–6
Rural Bus Partnership 113
service stability 51
staff security 55
timetables 51–2
Traffic Commissioners 42
transport policy 15
use by less affluent 26
vehicle excise duty 121

cars
    see also road traffic
    car clubs 148
    company 122
    crime 54
    households without 26
    New Deal for motorists 16, 46
    ownership 12
    second-hand 140

Central Rail Users' Consultative Committee 29
Channel Tunnel Rail Link 81
Civil Aviation Authority 88–9, 101
Cleaner Vehicles Task Force 139
climate change
    see also environment, pollution
    aviation emissions 125
    CO_2 emissions 11, 25, 32–4
    global warming 25
    greenhouse gas emissions 17, 25
    greenhouse gas targets 31, 32–4
    Kyoto conference 17, 32, 126
coaches 66
    drivers' hours 86
    road safety 85–6
coastguard see Maritime and Coastguard Agency
Commission for Integrated Transport 34, 36, 72, 92–3, 113, 119, 151
Community Transport Charter 147
competition, transport policy 27–8
congestion
    costs of 11
    problems of 10, 24
    road user charging 115–17
Countryside Commission 62–3
Countryside Traffic Measures Group 62
crime reduction 54–5
cycling
    healthy option 22
    interchange facilities 49–50
    local transport plans 39
    London Cycle Network 108
    National Cycle Network 39–40
    National Cycling Forum 39
    National Cycling Strategy 38–9
    priority routes 59
    safety 39
    schoolchildren 39

development
    appraisal 134–5
    plans 45, 50–1, 58, 126–7
devolution 102
disabled people
    passenger information 52
    pedestrians 38
    transport policy 56–8
drink-driving 83

economy, transport policy 24–5
economic instruments 119–23
enforcement
    agencies 130
    British Transport Police 131
    freight transport 132
    police organisation 131
    road traffic 129
    technology 129–30
    wheelclamping 131
England, regional transport strategies 102–4
environment
    see also air quality, climate change, pollution
    effects of transport 10
    environmental appraisal 133–4
    fiscal incentives 119–21
    fuel duty 119–21
    fuel standards 123
    global warming 25
    green travel incentives 122
    Greener Cities emissions checks 130
    Greener Motoring Guide 140
    marine clean-up 125
    noise pollution 135–6
    ports and shipping 124–5
    residential 61–2
    road freight 74
    rural 62–3
    taxation 119–22
    targets 32–4
    transport policy 16–17
    trunk roads 69–70
    vehicle standards 123
European Union
    action on integrated transport 92
    air quality 124
    awareness raising 149–50
    fuel standards 123
    greenhouse gas emissions 17
    greenhouse gas reduction 33, 35
    noise standards 135

INDEX

road safety measures 83
targets and standards 31
technology research and development 137
town centre traffic systems 61
Trans-European Networks 80–2
vehicle standards 123

fares
    buses 43, 48–9
    integrated transport 48–9
    railways 95–6
ferries, disabled people 58
Foresight Transport Panel 61
Foresight Vehicle Initiative 139
freight
    air 74
    axle weights 71–2
    CO_2 emissions 25
    coastal shipping 75–6
    Commission for Integrated Transport 72
    drivers' hours 86
    enforcement 132
    environmentally friendly lorries 74
    improving efficiency 70–3
    inland waterways 75–6
    lorry excise duties 71, 72, 121
    Quality Partnerships 18, 46, 73
    rail freight grants 100
    rail tonnage reduction 24
    railways 44–5
    Strategic Rail Authority 18
    sustainable distribution 70–6
    sustainable shipping 74–5
    Trans-European Networks 81–2
    transport policy 18
    unsuitable roads 73–4
fuel duty 120–3
fuel standards 122–3
funding 93–4

global warming *see* climate change
Going for Green campaign 150
Government Car Service 121
Greater London Authority 105–6
green transport plans 19, 142–4

green travel incentives 122
Greener Motoring Guide 140
group travel incentives 55

Health and Safety Commission 86–7, 89
health
    green transport plans 142–4
    relationship with transport 22–4
    transport impact assessment 132
Highways Agency
    development control 70
    maintenance role 100–1
    network operator 66–7
    Toolkit 65–6
hospitals 143
housing, planning 127
    quality residential environments 61–2

inclusive society
    building communities 147–8
    local transport plans 111–12
    New Deal 26–7, 35
    transport policy 17–18
integrated transport 37–89
    airport integration 76–80
    Commission for Integrated Transport 34, 36
    competition 27–8
    European Union 92
    fares and ticketing 48–9
    Greater Manchester pilot project 110
    local roads 58–60
    London 105–9
    monopolies 27–8
    passenger information 51–3
    personal safety 54–5
    physical interchange 49–51
    policy 13
    port integration 80–2
    publications 154
    regulation 27–8
    safety 89
    seamless journeys 47–8
    service stability 51
    taxis 53
    timetable co-ordination 51

town centres 60–1
Traffic Commissioners 51
trunk roads 63–70
Intelligent Transport Systems 138

journey planning 51
    driver information 68–9
journeys to school 12, 32, 145–7
    road safety 83
    Safe Routes to School project 146
    safety aspects 23, 38
journeys to work 142–4

Kyoto conference 17, 125
    targets 32–4

land use planning *see* planning
Leicester, road user charging 115
lighting, roads 69–70
local democracy 18–19
local transport plans
    airports 78–9
    buses 42
    cycling 39
    development control 70
    development plans 127
    disabled people 56
    fares and ticketing 48
    freight 73
    funding buses 112
    funding major schemes 114
    inclusive society 112–13
    interchange facilities 49
    local roads 58
    lorries 73
    motorcycles 47
    parking 117–18
    proposals for 3, 14–15, 111–12
    rail services 114
    railways 45
    road user charging 116
    rural environments 63
    seamless journeys 47
    targets 111
    taxis 53

    timetable co-ordination 51
    workplace parking charges 117
London
    bus lane enforcement 129
    congestion charging study 116
    Greater London Authority 105, 108
    integrated transport 105–9
    parking 109
    Transport for London 105
    travel information 109
    Underground 106
lorries
    see also freight
    axle weights 71–2
    CO_2 emissions 25
    drivers' facilities 68
    drivers' hours 86
    vehicle excise duty 71, 72, 120–1
    vehicle standards 123

Manchester
    airport 78
    Metrolink 45
    pilot integrated transport project 110
Maritime and Coastguard Agency 87–8, 89
minicabs 53
monopolies, transport policy 27–8
motorcycles 47
    road safety 85
motorists, New Deal 16, 46

National Air Traffic Services 89, 101–2
National public transport information system 52
noise
    airports 23, 25, 101
    effects of 135
    mitigation 136
    road traffic 27, 135–6
Northern Ireland, scope of White Paper 8

parking
    controls 59
    London 107
    non workplace 118
    workplace 117

INDEX

Passenger Transport Authorities 110
passenger information 51–3
pedestrians 37–8
    priority in town centres 60
    priority routes 59
    road safety 82–4
    World Squares for All 108
pensioners, bus passes 35, 112
planning
    appraisal 134
    development plans 45, 50–1, 58, 126–7
    design 37–8, 39, 50, 56, 58, 61, 128
    housing 127
    implementation 127–8
    planning guidance notes 126–7
    Regional Planning Guidance 58, 63, 70, 102–4
    transport policy 126–7
    trunk roads 103
police
    British Transport Police 131
    enforcement role 130
    objectives for safer transport 54
    organisation 130–1
pollution *see also* climate change, environment
    air quality 23, 35, 124–5
    CO_2 emissions 25
    in-car 11
    light 69–70
    local roads 114–15
    National Air Quality Strategy 31, 58, 124
    noise 23, 25, 27, 135–6
    problems of 11
    road traffic 23
    Royal Commission on Environmental Pollution 13, 25, 31, 34, 118, 156–7
    targets 32–4
ports *see* shipping
private hire vehicles 53
public transport
    see also buses, railways
    airport access 46
    CO_2 emissions 25
    disabled people 27, 56–8
    less affluent people 26

New Deal 15, 29, 42
personal security 27
rural communities 12

Rail Regulator 28, 29, 48, 51–2, 97–9
railways
    airport links 79
    bottlenecks 43–4
    Central Rail Users' Consultative Committee 29
    Channel Tunnel Rail Link 81
    disabled people 56–8
    easier access 56–8
    fares and ticketing 43, 48–9, 95–6
    franchise operators 96
    freight 44–5
    freight grants 100
    funding local services 114
    Greater Manchester Metrolink 45
    infrastructure investment 97–8
    interchange facilities 49–51
    investment 88–9
    land disposal 100
    light rail 45
    local 45
    local transport plans 45
    London Underground 106
    passenger accountability 96–7
    passenger information 51–3
    passenger services 43–4
    port access 80
    privatisation 28–9
    Rail Regulator 28, 29, 48, 51–2, 97–9
    Railway Inspectorate 86
    rolling stock leasing companies 98
    rural 149
    safety 86–7
    Secure Stations Scheme 55
    service improvements 96–7
    service stability 51
    staff security 55
    Strategic Rail Authority 15–16, 42, 43, 44, 79, 80, 94–5, 96–100, 103, 114
    timetable co-ordination 51
    Trans-European Networks 80–2
    transport policy 15–16

Regional Development Agencies 103, 104
Regional Planning Guidance 58, 63, 70, 102–4
Regional Traffic Control Centres 67
regional transport strategies 102–4
regulation, transport policy 27–8
road building, reduction in 10, 12
road safety 82–6
    buses 85–6
    coaches 85–6
    cycling 39
    drink-driving 83
    drivers' hours 86
    European Union 83
    improvements 82
    motorcycles 85
    Safer Routes to School 146
    speed policy 84
road traffic
    see also buses, cars, freight, lorries, trunk roads
    20mph zones 61
    breakdown recovery 67–8
    bus deregulation 28
    bus priority 59
    Clear Zones 61
    coaches 66
    congestion reduction 34
    driver information 59, 68–9
    enforcement 129
    forecasts 11
    Foresight Transport Panel 61
    freight 70–4
    green transport plans 142–4
    high occupancy vehicle lanes 58, 59
    journeys to work 142–4
    local congestion 114–15
    local management 59
    London traffic management 107
    motorcycles 47
    noise 23, 27, 135–6
    parking 59
    pollution 23
    priority routes 59
    Road User's Charter 16
    road building policy 10, 12, 63–7
    road safety 23
    road user charging 14, 115–16
    rural 62–3
    safety 82–6
    speed 84
    taxis 53
    traffic calming 38, 59
    urban traffic control 59
    wheelclamping 131–2
    workplace parking 117
Royal Commission on Environmental Pollution 13, 25, 31, 34, 119, 156–7
rural transport 12, 62–3, 147–8
    Rural Bus Partnership 113
    Rural Traffic Advisory Service 62
    Rural Transport Partnership scheme 112, 148

safety
    see also road safety
    air 88–9
    cycling 39
    integrated approach 89
    marine 87–8, 124
    personal 54–5
    railway 86–7
    road 82–6
school journeys 12, 32, 145–7
    road safety 83
    Safer Routes to School project 146
    safety aspects 23, 38
Scotland, scope of White Paper 8
seamless journeys 47–8
shipping
    coastal 75–6
    environmental impact 124
    inland waterways 75–6
    integrated ports 80–2
    marine clean-up 125
    MV *Derbyshire* 79
    pilotage 125
    safety 87–8, 124–5
    Sea Empress 125
    sustainable 74–5
    Thames 2000 project 76
    trust ports 102

INDEX

Strategic Rail Authority 15–16, 18, 42, 43, 44, 79, 80, 94–5, 96–100, 103, 114

taxis 53
    disabled people 56
technology
    enforcement 129–30
    integrated payment systems 49
    Intelligent Transport Systems 138
    passenger information 53
    research and development 137–8
    road user charging 116
    transport policy 30
teleworking 144–5
Thames 2000 project 76
Thames Gateway regeneration 104
ticketing
    integrated transport 48–9
    London Travelcard 48
town centres 60–1
Traffic Commissioners 42, 51
Trans-European Networks 65, 80–2
Transport for London 106
transport preferences 28
Transport: the way forward, Green Paper 12, 158–60
travel habits
    changing 19, 29–30, 114–18
    comparative prices 119
    economic instruments 119–20
    leisure travel 144
    shared responsibility 141–2
    shopping 144
    transport preferences 28
    travelling to work 142
TravelWise campaign 150
trunk roads
    see also road traffic
    development control 70
    driver information 68–9
    Highways Agency 66–7
    integrated network 63, 64–5
    investment 64, 100–1
    lighting 69–70
    local environment 69–70
    lorry drivers' facilities 68
    network 64–5, 161
    New Deal 63–70
    planning 103
    Regional Traffic Control Centres 67
    Road User's Charter 67
    road user charging 116
    use of 65–6

vehicle excise duty 120–1
vehicle standards 123

Wales, scope of the White Paper 8
walking *see* pedestrians
wheelclamping 131
women, transport policy 46

Printed in the UK for The Stationery Office Limited
on behalf of the Controller of Her Majesty's Stationery Office
Dd 5068056 8/98 13110 Job No. J0072938